31st Annual Meeting of the North American Membrane Society (NAMS 2022)

Membranes for Sustainability

Tempe, Arizona, USA and Online
14-18 May 2022

Editor:

Weiliang Bai

ISBN: 978-1-7138-5550-7

Printed from e-media with permission by:

Curran Associates, Inc.
57 Morehouse Lane
Red Hook, NY 12571

Some format issues inherent in the e-media version may also appear in this print version.

North American Membrane Society (NAMS)
University of Arkansas
Ralph E. Martin Department of Chemical Engineering
3202 Bell Engineering Center
Fayetteville, AR 72701-1201
USA

Phone: +1 479 575-7926
Fax: +1 479 575-7926

names@uark.edu

Additional copies of this publication are available from:

Curran Associates, Inc.
57 Morehouse Lane
Red Hook, NY 12571 USA
Phone: 845-758-0400
Fax: 845-758-2633
Email: curran@proceedings.com
Web: www.proceedings.com

North American Membrane Society

31ST ANNUAL MEETING

2022 HYBRID MAY 14TH – 18TH

TEMPE MISSION PALMS, TEMPE, AZ

MEMBRANES FOR SUSTAINABILITY

@2022NAMS

www.membranes.org/nams~2022/

Program Book

Conference chairs

Mary Laura Lind, Arizona State University

Kerri Hickenbottom, University of Arizona

Manish Kumar, The University of Texas at Austin

Vicki Karanikola, University of Arizona

Edited by

Weiliang Bai, The University of Texas at Austin

Acknowledgement

Image for program book cover from

Prof. Christopher G. Arges

Department of Chemical Engineering, Penn State University

North American Membrane Society

Table of Contents

Welcome to NAMS 2022

On behalf of the NAMS Board of Directors, we are excited to welcome you to NAMS 2022 at the celebrated Mission Palms Hotel and Conference Center in Tempe, Arizona!

Many thanks to our NAMS 2022 chairs Mary Laura Lind, Manish Kumar, Kerri Hickenbottom, and Vicki Karanikola for organizing the first fully hybrid meeting in NAMS history and putting together an exciting program focused on the theme of *Membranes for Enabling Sustainability*. We are most grateful for their hard work in service to NAMS and the membrane community. We also thank all members of the organizing committee, our generous sponsors, our hardworking staff, and program session chairs for helping to make NAMS 2022 a great success!

NAMS 2022 brings together participants from around the world and provides a platform for scientific exchanges on new ideas and the latest research in membrane science and technology. A comprehensive technical program is in place that includes three plenary talks from distinguished speakers, 180 talks, and an evening poster session. Among the sessions is the NAMS award session, in which the Student Fellowship and Young Membrane Scientist award winners will present their work.

NAMS 2022 also gives us an excellent opportunity to recognize the career-long success of society members. This year, we are honored to recognize three new NAMS Fellows for their important contributions in service to NAMS and their highly significant professional accomplishments in the membrane field. Congratulations to Yoram Cohen, Michael Guiver, and Andrew Zydney!

NAMS 2022 offers a number of exciting opportunities for networking among membraneologists from industry, government, and academia, as well as many students working in the membrane field. NAMS is dedicated to mentoring this next generation of membraneologists through its student program, industry-student mentorship program, and travel supplements to increase student participation at our meetings. Many of the students attending NAMS 2022 will present posters at the Monday evening poster session, which is sure to be a highlight of the meeting!

Have a great meeting, and don't forget to pause every now and then to enjoy all the beauty that this magnificent desert setting has to offer!

Scott Husson, NAMS President and Ryan Lively, NAMS Vice President

Message from the 2022 NAMS Conference Chairs

It is our pleasure to welcome you to the 31st Annual Meeting of the North American Membrane Society, and the first ever NAMS meeting to be held in a fully hybrid mode. It has been our honor to serve as co-chairs for this year's NAMS meeting given the challenges of the times. NAMS hosted the first fully online meeting in 2020, one of the first societies to rapidly adapt to the COVID crisis. This meeting has served as a model for how such events can bring expanded access to people with various backgrounds and was highlighted in a Nature Sustainability study. With the pandemic continuing to have an impact on our lives, and with travel restrictions still in place, we decided to run this meeting in the hybrid mode to combine the best of both formats and to provide expanded opportunities for participation.

During the conference, the plenary session, oral sessions, and the banquet live streamed, and with speakers and moderators both online and in person. We will use Whova as a platform for seamless integration and networking as well for streaming talks so make sure you download the app on your phone and bookmark it on your computer. We will have approximately 150 posters presented and 180 talks. These contributions are of the highest quality (as seen from the abstracts) and we are sure to have an engaging conference.

A very warm thank you to our dedicated oral and poster session chairs. Also special thanks to the NAMS 2022 Conference Advisory Committee and supporters (Glenn Lipscomb, Isabel Escobar) and NAMS board members and leadership (especially Jamie Hestekin, Christina Carbrello, Lucy Camacho, and Scott Husson) for all the input and support. We would also like to acknowledge our Fundraising chair JR Johnson and the Workshop Chair Marie-Eve Langevin as well as our sponsors who helped us raise substantial funds to support the conference and conference attendees.

We are truly excited to host you in Tempe and online. Hopefully you are excited to attend NAMS!

Kerri Hickenbottom, Vicky Karanikola, Manish Kumar, and Mary Laura Lind

NAMS 2022 Conference Co-Chairs

NAMS 2022 Anti-Harassment Policy

NAMS 2022 is dedicated to providing a harassment-free conference experience for everyone. We will not tolerate harassment of conference participants (including sexual harassment and assault) in any form. Conference participants violating these rules may be sanctioned or expelled from the conference without a refund at the discretion of the conference organizers. Harassment includes, but is not limited to: comments related to gender, gender identity and expression, sexual orientation, physical appearance, race, or religion; unwanted sexual attention; unwanted physical contact; threats of violence; intimidation; and stalking.

Reporting: If someone makes you or anyone else feel unsafe or unwelcome, please report it as soon as possible to one of the conference co-chairs. If you are being harassed by someone, notice that someone else is being harassed or have any other concerns, please contact one the conference co-chairs through the Whova App or in person

Enforcement: If a participant engages in harassing behavior, event organizers retain the right to take any actions to keep the event a welcoming environment for all participants. This includes warning the offender or expulsion from the conference with no refund. Event organizers may take action to redress anything designed to, or with the clear impact of, disrupting the event or creating a hostile environment for attendees. We expect participants to follow these rules at all event venues and event-related social activities.

In Memoriam

This year, membrane society was saddened by the passage of two great researchers, Dr. Srinivasa Sourirajan and Dr. Suk-Tak Hwang. Here are brief introduction and accomplishment by them. They and their wonderful work will be remembered by the society.

Dr. Srinivasa Sourirajan

Dr. Srinivasa Sourirajan, who is widely regarded as one of the inventors of the practical reverse osmosis (RO) membrane, passed away in Ottawa, Canada on 20thFebruary, 2022. He was born in a rural village in India and received his PhD in Chemistry. After moving to the US, he received an additional doctorate degree in Chemical Engineering at Yale University and then joined UCLA department of engineering as a research assistant. At UCLA, he invented the phase inversion process to synthesize anisotropic cellulose acetate membranes in collaboration with Dr. Sidney Loeb in 1960. Owing to the thin active skin layers, they achieved a desalination flux that was high enough to be commercially viable. This enabled large-scale industrialization of membrane separation processes and which is now a tool to address one of the critical challenges faced by global community–access to clean water. To expand the applications of membranes beyond water desalination, he moved to Ottawa and worked in National Research Council of Canada and extended his research to liquid and gas separations in industrial, medical, health care and environmental applications. Acknowledging his accomplishments, he was nominated for the Nobel prize in chemistry three times. He also received Honorary Doctorate Degree by the University of Ottawa, became a part of American Membrane Technology Association's Hall of Fame, and was nominated as one of top-achieving chemical engineers of the 20th century by the Canadian Chemical Engineering Conference.

Dr. Sun-Tak Hwang

Dr. Sun-Tak Hwang, a renowned membrane expert, passed away in Blue Ash, Ohio on March 6th this year. He was born near Daejeon, South Korea in 1935. He earned a bachelor's degree in chemical engineering from Seoul National University before coming to the University of Iowa to earn a master's degree (1962) and a PhD (1965), also in chemical engineering. Shortly afterwards, he joined the faculty at the University of Iowa. In 1982, he joined the University of Cincinnati to serve as a department head and helped start the Center of Excellence for Membrane Technology. He also played a key role in the founding of the North American Membrane Society in 1985. He was a leader in the membrane field and his research group conducted work in a variety of areas, including surface diffusion, gas and vapor separations, pervaporation, hollow-fiber membranes, membrane columns, and membrane casting. He also co-authored the book "Membranes in Separations" with Karl Kammermeyer in 1975. In recognition of his contributions to engineering, he was awarded the 1998 Ho-Am Prize in Engineering, which is awarded annually to one South Korean whose work in engineering benefits humanity. He was also inducted into the University of Iowa Distinguished Engineering Alumni Academy in 2003.

NAMS Fellows are recognized members of NAMS who have demonstrated ongoing excellence in membrane engineering and practice. Election as Fellow is in recognition of both "service to NAMS" and "highly significant professional accomplishment in the membrane field."

Yoram Cohen

Dr. Yoram Cohen is a distinguished professor of Chemical and Biomolecular Engineering at the University Of California, Los Angeles (UCLA) and a faculty member at the UCLA institute of the environment and sustainability. Dr. Cohen's work on membrane surface nano-structuring culminated in the development of high performance spiral-wound RO membranes. His pioneering work on the basic principles of energy utilization in RO systems paved the way for RO energy consumption minimization.

Dr. Cohen has a sustained record of 40 years of contributions to membrane science and technology. He published nearly 270 scholarly publications, edited 7 books, served as Guest Editor of 3 special journal issues. Dr. Cohen has been a member of NAMS for 25 years and co-chair the 2016 NAMS Annual Meeting held in Bellevue, Washington.

Michael Guiver

Dr. Michael Guiver serves as National 100-Plan Foreign Experts Professor at Tianjin University, China. Dr. Guiver has worked in a wide variety of membrane research, particularly in the areas of polymer synthesis, membrane formation, fuel cells, gas- and vapor separations. From 1987-2014, he was a scientist at the National Research Council Canada. In 2009-2013, he was a visiting professor at the Department of Energy Engineering, Hanyang University, Korea under the "World Class University" program.

Dr. Guiver has been a member of NAMS for 22 years and has provided an outstanding service to the society. He severed as on the Advisor Board and Work committee of ICOM 2008 that was organized by NAMS. Dr. Guiver served NAMS on the Board of Directors from 2006 - 2009, and he also co-organized the 1996 NAMS meeting in Ottawa, Canada.

Andrew Zydney

Dr. Andrew Zydney is the Bayard D. Kunkle Chair and the Professor of Chemical Engineering at Penn State University, PA. Dr. Zydney's research has addressed wide ranging problems in membrane science and technology with a specific focus on applications of membranes in bioprocessing. He developed the shear-induced diffusion model for the filtrate flux in particle microfiltration. He is currently pioneering the development of membrane systems for continuous biomanufacturing of monoclonal antibodies.

Dr. Zydney was elected to the Board of Directors of NAMS in 1998, serving on the Board for the next 8 years. He co-taught a NAMS Workshop on Membrane Fouling in 1994, 1995, 1998 and 2000. Dr. Zydney was also invited to serve as an ad hoc Chair of the NAMS Membership Committee in 1995 and co-organized two highly successful NAMS meetings: Baltimore in 1997 and Pittsburgh in 2019.

Award Session – Room Palm AD, 9:30 am -12:30 pm, Wednesday, May 18th

Student Fellowship Award

The NAMS Student Fellowship Awards are presented annually to outstanding graduate students in the area of membrane science and technology.

Adam Uliana
University of California Berkeley
11:00 am, Wednesday
Room Palm AD

Katherine Mizrahi Rodriguez
Massachusetts Institute of Technology
10:30 am, Wednesday
Room Palm AD

Ryan DuChanois
Yale University
12:00 pm, Wednesday
Room Palm AD

Young Membrane Scientist Award

The NAMS Young Membrane Scientist Awards are awarded to those individuals who are within five years of completing their Ph.D. degree and have already proven to be outstanding membrane scientists in the area of membrane science and technology.

Luis Francisco Villalobos
Yale University
11:30 am, Wednesday
Room Palm AD

Anthony Straub
University of Colorado
10:00 am, Wednesday
Room Palm AD

Christine Duval
Case Western Reverse University
9:30 am, Wednesday
Room Palm AD

➢ PROGRAM-AT-A-GLANCE

NAMS 2022 Meeting

Date	Time (*MST)	Event	ROOM
Saturday May 14	9:00 AM – 5:00 PM	Workshop I: Measurement Methods for Membranes	Abbey North
		Workshop II: Membranes for Water Treatment Applications	Abbey South
Sunday May 15	9:00 AM - 5:00 PM	Workshop III: Membranes for Gas Separations	Abbey South
	2: 00 PM - 5:00 PM	NAMS Student Workshop	Ballroom
	2:00 PM – 6:00 PM	Registration	Lobby
	7:00 PM - 9:00 PM	Welcome Reception	Courtyard
Monday May 16	7:00 AM – 8:00 AM	Registration	Lobby
	8:00 AM - 9:00 AM	Plenary Lecture 1	Ballroom
	9:00 AM – 9:30 AM	Coffee Break	Ballroom Foyer
	9:30 AM – 12:30 PM	Parallel Technical Sessions I	See Session Details
	12:30 PM – 2:00 PM	Lunch with Legends	Culinary Dropout
	2:00 PM - 5:00 PM	Parallel Technical Sessions II	See Session Details
	7:00 PM – 10:00 PM	Poster Session	Ballroom
Tuesday May 17	7:00 AM – 8:00 AM	Registration	Lobby
	8:00 AM – 9:00 AM	Plenary Lecture II	Ballroom
	9:00 AM – 9:30 AM	Coffee Break	Ballroom Foyer
	9:30 AM – 12:30 PM	Parallel Technical Sessions III	See Session Details
	12:30 PM - 2:00 PM	Lunch Break	
	2:00 PM – 5:00 PM	Parallel Technical Sessions IV	See Session Details
	5:30 PM – 6:30 PM	NAMS Business Meeting	Abbey South
	6:30 PM - 7:30 PM	Closing and Award	Ballroom
	7:30 PM - 11:00 PM	Banquet Dinner	Courtyard
Wednesday May 18	7:00 AM – 8:00 AM	Registration	Lobby
	8:00 AM – 9:00 AM	Plenary Lecture III	Ballroom
	9:00 AM – 9:30 AM	Coffee Break	Ballroom Foyer
	9:30 AM – 12:30 PM	Parallel Technical Sessions V	See Session Details
	12:30 PM – 2:00 PM	Lunch Break	
	2:00 PM – 5:00 PM	Parallel Technical Sessions VI	See Session Details

*MST – US Mountain Standard time, UK +8 hrs, Germany +9 hrs, Saudi Arabia (in Riyadh) +10 hrs, Inda (IST) +12.5 hrs, Singapore +15 hrs, China +15 hrs, South Korea +16 hrs, Japan +16 hrs.

Date	Time (*MST)	Event	Room
Monday May 16	8:00 AM - 9:00 AM	**Plenary Session 1**	**Ballroom**
	9:00 AM – 9:30 AM	**Break**	
	9:30 AM – 12:30 PM (Parallel Oral Sessions)	**Organic Solvent Separations I**	**Palm AD**
		Carbon Capture I	**Palm CF**
		Emerging Materials for Liquid Separations I	**Palm BE**
		Membrane Synthesis and Casting I	**Abbey North**
		Inorganic Materials	**Abbey South**
	12:30 PM – 2:00 PM	**Lunch Break**	
	2:00 PM - 5:00 PM (Parallel Oral Sessions)	**Organic Solvent Separations II**	**Palm AD**
		Carbon Capture II	**Palm CF**
		Emerging Materials for Liquid Separations II	**Palm BE**
		Membrane Synthesis and Casting II	**Abbey North**
		Catalytic and Responsive Membranes	**Abbey South**
	7:00 PM – 10:00 PM	**Poster Session**	**Ballroom**
Tuesday May 17	8:00 AM – 9:00 AM	**Plenary Session II**	**Ballroom**
	9:00 AM – 9:30 AM	**Break**	
	9:30 AM – 12:30 PM (Parallel Oral Sessions)	**Polymeric and Mixed-Materials – Gas Separations I**	**Palm AD**
		Bioinspired and Biomimetic Materials	**Palm BE**
		Membranes for Electrochemical Applications I	**Palm CF**
		High Salinity Streams, Brine Mineralization and ZLD	**Abbey South**
		Process Intensification and Integration	**Abbey North**
	12:30 PM - 2:00 PM	**Lunch Break**	
	2:00 PM – 5:00 PM (Parallel Oral Sessions)	**Polymeric and Mixed-Materials – Gas Separations II**	**Palm AD**
		Downstream Bioprocessing	**Palm BE**
		Membranes for Electrochemical Applications II	**Palm CF**
		Membrane Characterization	**Abbey South**
		Molecular and Process Modeling	**Abbey North**
Wednesday May 18	8:00 AM – 9:00 AM	**Plenary Session III**	**Ballroom**
	9:00 AM – 9:30 AM	**Break**	
	9:30 AM – 12:30 PM (Parallel Oral Sessions)	**Award Session**	**Palm AD**
		Membrane Fouling I	**Palm CF**
		Membrane Distillation and Pervaporation	**Abbey North**
		Industrial Applications	**Palm BE**
		Materials for Electrochemical Applications	**Abbey South**
	12:30 PM – 2:00 PM	**Lunch Break**	
	2:00 PM – 5:00 PM (Parallel Oral Sessions)	**Seawater Desal and Osmotic Process**	**Palm BE**
		Membrane Fouling II	**Palm CF**
		Contaminant Removal from Water	**Palm AD**
		Water Reuse	**Abbey North**
		Module Modeling and Design	**Abbey South**

*MST – US Mountain Standard time, UK +8 hrs, Germany +9 hrs, Saudi Arabia (in Riyadh) +10 hrs, Inda (IST) +12.5 hrs, Singapore +15 hrs, China +15 hrs, South Korea +16 hrs, Japan +16 hrs.

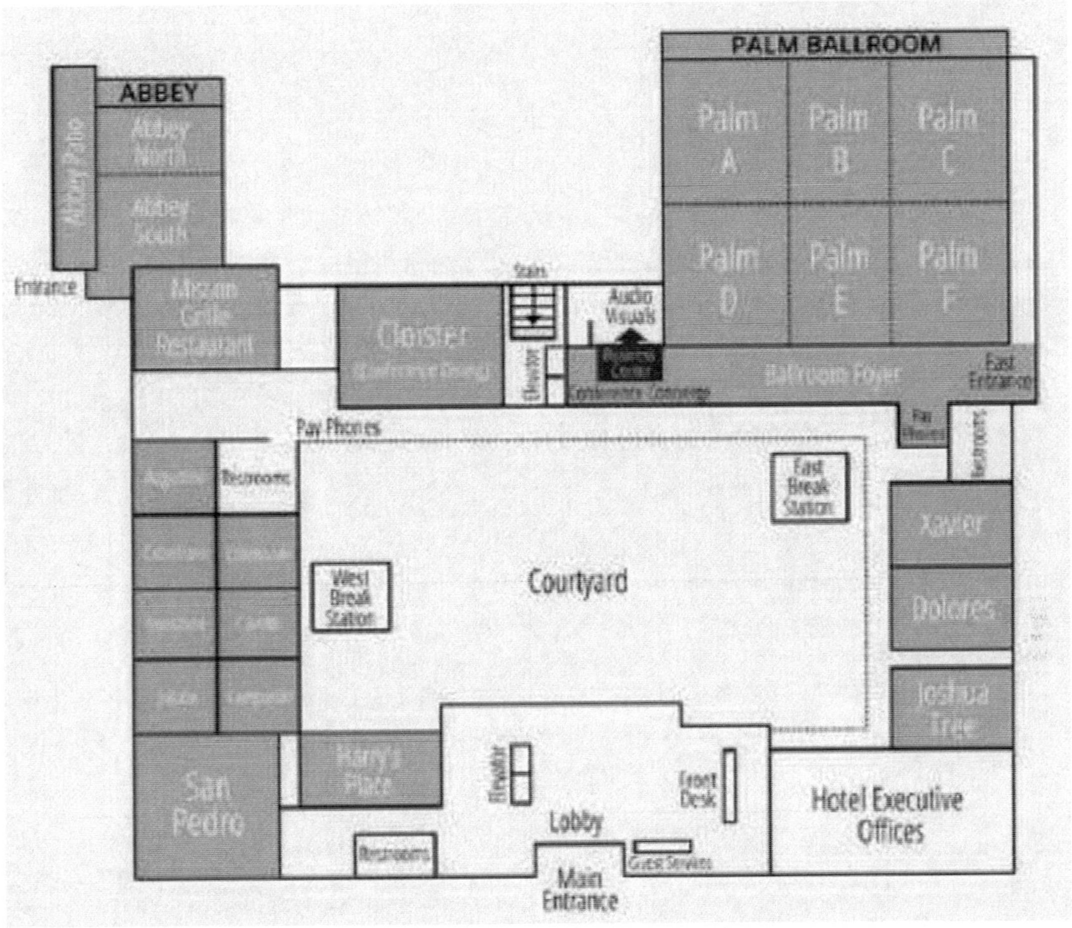
PALM BALLROOM
Palm A
Palm B
Palm C
Palm D
Palm E
Palm F
ABBEY
Abbey North
Abbey South
Abbey Patio
Entrance
Mission Grille Restaurant
Cloister
Stairs
Audio Visuals
Elevator
Ballroom Foyer
East Entrance
Pay Phones
Restrooms
Pay Phones
East Break Station
Restrooms
Xavier
Dolores
Joshua Tree
West Break Station
Courtyard
Harry's Place
San Pedro
Elevator
Front Desk
Lobby
Restrooms
Guest Services
Main Entrance
Hotel Executive Offices

8:00 AM	**Plenary Session I (Ballroom)**				
Parallel Sessions	**Room Palm AD** **Organic Solvent Separations I**	**Room Palm CF** **Carbon Capture I**	**Room Palm BE** **Emerging Materials for Liquid Separations I**	**Room Abbey North** **Membrane Synthesis and Casting I**	**Room Abbey South** **Inorganic Materials**
Chairs	**Michele Galizia** mgalizia@ou.edu **Steve White** steve.white@mtrinc.com **Neel Rangnekar** neel.d.rangnekar@exxonmobil.com	**Winston Ho** ho.192@osu.edu **David Hopkinson** David.Hopkinson@netl.doe.gov **Katherine Hornbostel** hornbostel@pitt.edu **Chen Zhang** czhang71@umd.edu	**Oishi Sanyal,** oishi.sanyal@mail.wvu.edu **Baoxia Mi** mib@berkeley.edu **Beza Getachew** bezawit.getachew@rice.edu	**Rachel Dorin** rachel.dorin@teraporetech.com **Christine E. Duval** ced84@case.edu **Abhishek Roy** abhishek.roy@nrel.gov **Reza Foudazi** rfoudazi@ou.edu	**Luis Francisco Villalobos** luisfrancisco.villalobos@yale.edu **Jay Kniep** jay.kniep@mtrinc.com **Minghui Qiu** qiumh_1201@njtech.edu.cn **David Bergsman** dbergs@uw.edu
99:30 AM	Oral 1 - From Renewable Building Blocks To Biodegradable Polymer Membranes With Solvent-Resistant Properties Gyorgy Szekely (King Abdullah University Of Science And Technology)	Oral 7 - Thin Film Composite Membranes With Novel Quaternary Ammonium Functionalized Poly(Arylene Ether Sulfone) Copolymers For Direct Air Capture Hoda Shokrollahzadeh Behbahani (Arizona State University)	Oral 13 - Separation Of Trace Organics From Aqueous Streams Using Carbonaceous Membranes Haley D White (Georgia Institute Of Technology)	Oral 19 - Microstructure Evolution During Nonsolvent-Induced Phase Separation: A Joint Experimental-Computational Investigation Into Membrane Formation Alexander Bridge (The University Of Texas At Austin)	Oral 25 - High-Pressure CO_2 Permeation Properties Of Ceramic-Carbonate Dual-Phase Membranes Jerry Lin (Arizona State University)
10:00 AM	Oral 2 - Towards Wired-Lab Organic Solvent Nanofiltration: Is Machine Learning Viable? Gergo Ignacz (King Abdullah University Of Science And Technology)	Oral 8 - Integrated Facilitated Transport Membrane Modules For Highly Selective Syngas Purification And Carbon Capture Yang Han (The Ohio State University)	Oral 14 - Cucurbit[N]Uril-Based Nanofiltration Membranes For Effective Molecular Separation Shi-Peng Sun (Nanjing Tech University)	Oral 20 - Automated Stepwise Synthesis Of Poly(Acrylonitrile)-Supported Oligoamide Membranes And Their Dye–Salt Separation Ron Kasher (Ben-Gurion University Of The Negev)	Oral 26 - Metal Ions-Doped Ultrahigh Flux Composite Graphene Oxide/Carbon Nanotubes (GO-SWCNT) Membranes For Efficient H_2 Separation Karel Friess (University Of Chemistry And Technology Prague)
10:30 AM	Oral 3 - Tuning Polyimide Thin Film Composite Membranes For Organic Solvent Reverse Osmosis Separations Via Boc Protected Amine Solid-State Crosslinking Yacine Feliachi (Georgia Institute Of Technology)	Oral 9 - Thin-Film Composite Membranes Based On Hyperbranched Poly(Ethylene Oxide) For CO_2/N_2 Separation Gengyi Zhang (The State University Of New York At Buffalo)	Oral 15 - Nanoporous Multilayer Graphene Membrane For Precise And Stable Forward Osmosis Desalination Claudio A Ruiz Torres (The University Of Toronto)	Oral 21 - Microporous Polymer Membranes: A New Approach To Simultaneously Reducing Physical Aging And Swelling While Enhancing Permeability Michele Galizia (University Of Oklahoma)	Oral 27 - Network Nanostructured ZIF-8 To Enable Percolation And Enhanced Gas Transport Hyunhee Lee (Massachusetts Institute Of Technology)
11:00 AM	Oral 4 - Organic Solvent Stable Graft Modified Polybenzimidazole Membranes With Customisable Rejection Profiles Adam E Oxley (Imperial College London)	Oral 10 - Polyvinylpyridine-Based Catalytic Membranes For Integrated CO2 Capture And Conversion: Structure-Function Relationships Casey O'Brien (University Of Notre Dame)	Oral 16 - High Flux Vapor-Gap Membranes For Reverse Osmosis Via Informed Design Of Membrane Pore Structures Elizabeth A Hjelvik (University Of Colorado Boulder)	Oral 22 - Crosslinked Triptycene-Containing Polybenzoxazole Membranes With Model Network Structures For High Performance Gas Separations Ziwei Dai (University Of Notre Dame)	Oral 28 - High-Performance CO_2- And O_2-Sieving By Controlled Etching Of Graphene Shiqi Huang (Ecole Polytechnique Fédérale De Lausanne)
11:30 AM	Oral 5 - Precision Molecular Sieving With MOF And COF Membranes In Organic Solvents Rifan Hardian (King Abdullah University Of Science And Technology)	Oral 11 - Effect Of Free Volume Modification (FVM) On The Physical Aging Behavior Of Amine-Functionalized PIM-1 Membranes For Gas Separation Taigyu Joo (Massachusetts Institute Of Technology)	Oral 17 - Preparation Of Functionalized Porous Poly (Ether Ether Ketone), PEEK, Materials And Hollow Fiber Membranes Ben Bikson (Avanpore)	Oral 23 - Integrated Roll-To-Roll (R2R) Fabrication Of Graphene Oxide Nanofiltration Membranes For Applications In Extreme Operating Environments Amirsalar Rabbani Esfahani (Georgia Institute Of Technology)	Oral 29 - Tunable Organosilica Membranes For High-Temperature H_2/CO_2 Separation Synthesized By Rapid Oxygen Plasma Treatment Of Polysiloxane Vinh Bui (University At Buffalo)
12:00 PM	Oral 6 - Engineering Organic Solvent Reverse Osmosis Performance Of Vapor Phase Infiltrated Organic-Inorganic Hybrid Membranes Yi Ren (Georgia Institute Of Technology)	Oral 12 - The Effect Of Channel Height Variation On CO_2 Separation Performance In A Flat Sheet Membrane Joanna Rivero (University Of Pittsburgh)	Oral 18 - Polyethyleneimine-Based Surface Treatment For Facile Synthesis Of Conductive Membranes: Active Material Doesn't Matter Wei Zhang (University Of North Carolina At Chapel Hill)	Oral 24 - Super-Hydrophobic MFI Zeolite Hollow Fiber Membranes For Alcohol/Water Separation Jerry Lin (Arizona State University)	Oral 30 - Synthesis Of ZIF Membranes By A Vapor-Phase Seeding Method For Olefin/Paraffin Separation Zhiqin Qiang (University Of Wisconsin-Milwaukee)

Monday Afternoon

Parallel Sessions	Room Palm AD Organic Solvent Separations II	Room Palm CF Carbon Capture II	Room Palm BE Emerging Materials for Liquid Separations II	Room Abbey North Membrane Synthesis and Casting II	Room Abbey South Catalytic and Responsive Membranes
Chairs	**Michele Galizia** mgalizia@ou.edu **Steve White** steve.white@mtrinc.com **Neel Ragnekar** neel.d.rangnekar@exxonmobil.com	**Winston Ho** ho.192@osu.edu **David Hopkinson** David.Hopkinson@netl.doe.gov **Katherine Hornbostel** hornbostel@pitt.edu **Chen Zhang** czhang71@umd.edu	**Oishi Sanyal,** oishi.sanyal@mail.wvu.edu **Baoxia Mi** mib@berkeley.edu **Beza Getachew** bezawit.getachew@rice.edu	**Rachel Dorin** rachel.dorin@teraporetech.com **Christine E. Duval** ced84@case.edu **Abhishek Roy** abhishek.roy@nrel.gov **Reza Foudazi** rfoudazi@ou.edu	**Miao Yu** myu9@buffalo.edu **Ayse Asatekin** ayse.asatekin@tufts.edu **Simona Liguori** sliguori@clarkson.edu **Volodymyr Tarabara** tarabara@egr.msu.edu
22:00 PM	Oral 31 - Interfacial Polymerization With Hydrophobic Amines Provides Ultrathin Polyamide Nanofilms For Hydrocarbon Separation Neel Rangnekar (Exxonmobil)	Oral 37 - Direct Ocean Capture Using Membrane Contactors Austin R Lieber (University Of Pittsburgh)	Oral 43 - Bioinspired Cross-Linked Iron Doped Reduced Graphene Oxide (rGO) Membrane For Ultrafiltration Of Organic Micropollutants Charles-François De Lannoy (Mcmaster University)	Oral 49 - Thin Zeolitic-Imidazolate Framework ZIF-8 Membranes Supported On ZnO-Deposited Substrates For Propylene/Propane Separation Anil Ronte (Oklahoma State University)	Oral 55 - Synthesis Of Reactive Membranes For Capture And Degradation Of Water And Air Pollutants Rollie G Mills (University Of Kentucky)
2:30 PM	Oral 32 - A Smart And Responsive Crystalline Porous Organic Cage Membrane With Switchable Pore Apertures For Graded Molecular Sieving Zhiwei Jiang (Queen Mary University Of London)	Oral 38 - Highly Permeable Substrates Prepared By Vapor-Induced Phase Separation For CO_2-Selective Composite Membranes Ruizhi Pang (The Ohio State University)	Oral 44 - Functional Nanofibrous Membranes Modified With B-Cyclodextrins For Improved Surface Characteristics And Filtration Performance Saranya Rameshkumar (Trinity College Dublin)	Oral 50 - 3D Printing Of Thin Film Composite Membranes For Nanofiltration And Beyond Jeffrey R Mccutcheon (University Of Connecticut)	Oral 56 - Ionic Strength And pH Responsive UF And RO Membranes Yoram Cohen (University Of California, Los Angeles)
3:00 PM	Oral 33 - Solution Processable Polytriazoles From Spirocyclic Monomers For Membrane-Based Hydrocarbon Separations Nicholas C Bruno (Georgia Institute Of Technology)	Oral 39 - Techno-Economic Assessment Of Postcombustion Capture Processes Using High-Performance Nanoporous Graphene Membranes Marina Micari (Ecole Polytechnique Fédérale De Lausanne)	Oral 45 - Cross-Linkable/ Self-Assembling Terpolymer Membranes For Brackish Water Desalination Samuel J Lounder (Tufts University)	Oral 51 - Advanced Ionic Polymers For High-Performance Gas Separation Membranes Sudhir Ravula (University Of Alabama)	Oral 57 - Catalyst-Free Water-Gas Shift Reaction In Ceramic-Carbonate Dual-Phase Membrane Reactors At High Pressures – Effect Of Side Reaction Jerry Lin (Arizona State University)
3:30 PM	Oral 34 - Organometallic Hybrid Nanofilms With Highly Interconnected Pore Architecture For Ultrafast Solvent Transport With Precise Molecular Separation Bratin Sengupta (University At Buffalo)	Oral 40 - Critical Aspects Of High-Pressure CO_2-Induced Plasticization In Polyimide Membranes Menno Houben (Eindhoven University Of Technology)	Oral 46 - Including Non-Binding 'Spacer' Monomers In Polyprotic Polymeric Ligands Impacts Ligand-Ion Affinity For Lanthanum Priyanka Suresh (Case Western Reserve University)	Oral 52 - Confined Thermally Induced Phase Separation Under Micromolding – Fabrication Of Patterned Membranes With Enhanced Pore Connectivity And High Salt-Scaling Resistance Shouhong Fan (University Of Colorado At Boulder)	Oral 58 Catalytic Membranes And The Role Of Biohydrogen For Reductive Chloro-Organics Detoxification From Water Dibakar Bhattacharyya (University Of Kentucky)
4:00 PM	Oral 35 - Effect Of Xylene Activity And Crystal Orientation Of MFI- Zeolite Membranes On Separation Of Xylene Isomers Jerry Lin (Arizona State University)	Oral 41 - In Situ Synergistic Growth Of Crystalline And Polymer-Incorporated Amorphous ZIF-8 In Polybenzimidazole Achieving Hierarchical Nanostructures For H_2/CO_2 Separation Leiqing Hu (University At Buffalo)	Oral 47 - Enhancing The Salt Rejections Of Covalent Organic Framework Membranes By Creating Dual-Layer Membrane Structure Miguel Jaimes (University Of Wisconsin-Milwaukee)	Oral 53 - Sustainability In Membrane Separations: Integration Of Eco-Friendly Materials And Scaled-Up Fabrication Of Polymeric Membranes For Water Purification David Lu (University Of Kentucky)	Oral 59 - Catalytic Membrane Reactor For Conversion Of Waste Biomass To Chemical Intermediates Ranil Wickramasinghe (University Of Arkansas)
4:30 PM	Oral 36 - Data-Driven Development Of Polymer Membrane Materials For Complex Mixture Separations Youngjoo Lee (Georgia Institute Of Technology)	Oral 42 - Mitigated Carrier Saturation Of Facilitated Transport Membranes For Decarbonizing Dilute CO_2 Sources Yang Han (The Ohio State University)	Oral 48 - Polyampholyte Copolymer Self-Assembly For Fouling Resistant And Easily Tunable Membranes Luca Mazzaferro (Tufts University)	Oral 54 - Polyamide Thickness And Morphology Features Of Thin-Film Composite Membranes Depend On The Extent Of Restriction Of The Amine Monomer Supply During Interfacial Polymerization Mikayla D Armstrong (University Of North Carolina At Chapel Hill)	Oral 60 - Trends, Mechanisms, And Opportunities In Using Electrically Responsive Membranes Charles-François De Lannoy (Mcmaster University)
7:00 PM	Poster Session (Ballroom)				

Tuesday Morning

8:00 AM	Plenary Session II (Ballroom)				
Parallel Sessions	**Room Palm AD** Polymeric and Mixed-Materials – Gas Separation I	**Room Palm BE** Bioinspired and Biomimetic Materials	**Room Palm CF** Membranes for Electrochemical Applications I	**Room Abbey South** High Salinity Streams, Brine Minimization and ZLD	**Room Abbey North** Process Intensification and Integration
Chairs	**Benjamin J. Sundell** benjaminjsundell@gmail.com **Zachary Smith** zpsmith@mit.edu **Xiaoli Ma** ma26@uwm.edu **Raj Singh°** rsingh@lanl.gov	**Yuexiao Shen** Yuexiao.Shen@ttu.edu **Cassandra Porter** cjp0084@auburn.edu	**Hee Jeung Oh** hjoh@psu.edu **Piran R Kidambi** piran.kidambi@vanderbilt.edu	**Kerri Hickenbottom** klh15@email.arizona.edu **Jonathan A. Brant** jbrant1@uwyo.edu	**Henry Jonathan Tanudjaja** henry.tanudjaja@ntu.edu.sg **Marie-Eve Langevin** MELangevin@ameridia.com **Mahdi Malmali** mahdi.malmali@ttu.edu **Albert S. Kim** AlbertSK@hawaii.edu
9:30 AM	Oral 61 - Unexpected Size-Controlled Vapor Sorption In Glassy Membranes Exhibiting Configurational Free Volumeconfigurational Free Volume William J. Box (University Of Oklahoma)	Oral 67 - Artificial Water Channels-Towards Biomimetic Membranes For Desalination Mihail Barboiu (Institut Europeen Des Membranes)	Oral 73 - Deconstructing Proton Transport Through Atomically Thin Monolayer CVD Graphene Membranes Piran Kidambi (Vanderbilt University)	Oral 79 - High Pressure Reverse Osmosis Membrane Modeling And Characterization Jeffrey R Mccutcheon (University Of Connecticut)	Oral 85 - Assessment Of A Hybrid Forward Osmosis-Freeze Crystallization Process For Liquid Mining Of Lithium: Experimental Study And Numerical Simulation Afshin Amani (Université Laval)
10:00 AM	Oral 62 - Microporous Polymers With Tailored Cavities By Direct Fluorination For Small Gas Molecule Separations Wonhee Lee (Georgia Institute Of Technology)	Oral 68 - Quantification Of Salt And Proton Rejection In A Flourofoldamer Based Artificial Water Channel Laxmicharan Samineni (University Of Texas At Austin)	Oral 74 - Investigation Of Porous PVDF Membranes Behavior During Metal-Assisted Chemical Imprinting Bruno Azeredo (Arizona State University)	Oral 80 - Selecting Fouling And Scaling Mitigation Strategies For Membrane Distillation Applications With Water Reuse Reverse Osmosis Concentrate Luke Presson (University Of Arizona)	Oral 86 - Thermal Storage Management Integrated With Solar Driven Membrane Distillation For Process Intensification Kerri Hickenbottom (University Of Arizona)
10:30 AM	Oral 63 - Pims At Nanoscale Wojciech Ogieglo (KAUST)	Oral 69 - Fast Water Transport Through Biomimetic Reverse Osmosis Membranes Embedded With Peptide-Attached (Pr)-Pillar[5]Arenes Water Channels Yu Jie Lim (Nanyang Technological University)	Oral 75 - Dialysate Regeneration Via Photooxidation And FO Membrane Protection-Loop To Enable Portable Kidney Dialysis Bruce Hinds (University Of Washington)	Oral 81 - Highly Charged Ion-Exchange Membranes For Treatment Of Brines Via Electrodialysis Carolina Espinoza (University Of Michigan)	Oral 87 - Reverse Osmosis Vs. Electrodialysis: Comparing The Cost Of Brackish Water Desalination Sohum Patel (Yale University)
11:00 AM	Oral 64 - Polyformamidine As Fixed-Site Carrier For CO_2/N_2 Separation Membranes Jingying Hu (The Ohio State University)	Oral 70 - Preparation Of Capillary Flow Membranes From Wood Veneers With Advanced Electron And Ion Microscopy Characterization Nadezda Prochukhan (Trinity College Dublin)	Oral 76 - Electrochemical Oxygen Pumps For Mobile Medical Oxygen Ralph A Bauer (Global Research And Development Inc.)	Oral 82 - Characterizing Low-Salt-Rejection RO Membranes With The Kedem-Katchalsky Model Abdessamad Belgada (Yale University)	Oral 88 - Biocidal Effects In Microwave Induced Membrane Distillation Indrani Gupta (New Jersey Institute Of Technology)
11:30 AM	Oral 65 - Scalable One-Step Fabrication Of MOF-Based Asymmetric Mixed-Matrix Membranes With Exceptionally High Propylene/Propane Separation Performance Yinying Hua (Texas A&M University)	Oral 71 - Skin Like Biomimetic Membranes For Breathable Protective Fabrics Hyeonji Oh (University Of Texas At Austin)	Oral 77 - Liquid Separations Membranes Made Of Functionalized Poly-Ether Ether Ketone Blended With Polysulfone Abelline K Fionah (University Of Kentucky)	Oral 83 - Screening For Scaling Resistant Desalination Membranes John Pellegrino (University Of Colorado Boulder)	Oral 89 - Automated And Field-Deployable Evaluation Of Microfiltration/Ultrafiltration Fouling Weiming Qi (Clemson University)
12:00 PM	Oral 66 - Effect Of PPE Decontamination Processes On The Filtration Efficiency Of Porous Polysulfone Flat Sheet Membrane Ebuka Ogbuoji (University Of Kentucky)	Oral 72 - A TiO_2-Au Janus Membrane Has Potential To Enable Self-Pumping Flow By Light. Yuhang Fang (Purdue University)	Oral 78 - Lithium Brine Purification And Electrochemical Conversion Via Selective Bipolar Membrane Electrodialysis Kevin Reimund (University Of Texas At Austin)	Oral 84 - Optimizing Pore Size And Charge Of Polyamide Nanofiltration Membrane For Lithium Enrichment From Salt Lake Brine Juan Zhai (Texas Tech University)	Oral 90 - Scale-Up Of Membrane Distillation For Treating Produced Waters From Oil And Gas Industry. Ritesh Pawar (University Of Pittsburgh)

Parallel Sessions	Room Palm AD PPolymeric and MixedMaterials – Gas Separation II	Room Palm BE Downstream Bioprocessing	Room Palm CF Membranes for Electrochemical Applications II	Room Abbey South Membrane Characterization	Room Abbey North Molecular and Process Modeling
Chairs	**Benjamin J. Sundell** benjaminjsundell@gmail.com **Zachary Smith** zpsmith@mit.edu **Xiaoli Ma** ma26@uwm.edu **Raj Singh** rsingh@lanl.gov	**Onur Kas** Onur.Kas@ucb.com **James McGrath** jmcgrath@bme.rochester.edu	**Hee Jeung Oh** hjoh@psu.edu **Piran R Kidambi** piran.kidambi@vanderbilt.edu	**Weiyi Li** liwy3@sustech.edu.cn **Santiago Romero** Santiago@ed.ac.uk **Yang Liu** Yang.Liu@aramcoamericas.com	**David M. Warsinger** david.warsinger@gmail.com **Christina Carbrello** christina.carbrello@milliporesigma.com **Shihong Lin** shihong.lin@vanderbilt.edu
2:00 PM	Oral 91 - Reduction Of Physical Aging In Copper Nanoparticle Pillared CMSMs Derived From A Polyimide Precursor Incorporating The Bis(Phenyl)Fluorene-Based Cardo Moiety And Uniformly Dispersed Metal Organic Polyhedra-18 (MOP-18) Masoumeh Tajik Asl (The University Of Texas At Dallas)	Oral 97 - Membrane Adsorber Purification Of pDNA Herb Lutz (Milliporesigma)	Oral 103 - Separation Of Rare Earth Elements From Simulated Geothermal Water Using Novel Electrodialysis Metathesis Process Lucy M Camacho (Texas A&M University-Kingsville)	Oral 109 - Multi-Lab Study On The Pure-Gas Permeation Of Commercial Polysulfone (PSF) Membranes: Measurement Standards And Best Practices Katherine Mizrahi Rodriguez (Massachusetts Institute Of Technology)	Oral 115 - The Impact Of Ion-Ion Correlated Motion On Salt Transport In Solvated Ion Exchange Membranes Nico Marioni (The University Of Texas At Austin)
2:30 PM	Oral 92 - Novel Graphene Oxide-Based Membrane Structure For A Highly Effective Breathable Barrier For Toxic Vapors And Chemical Warfare Agents Kamalesh K Sirkar (New Jersey Institute Of Technology)	Oral 98 - Predicting The Transport Of Soft Droplets In Porous Media From Measurable Emergent Properties John Pellegrino (University Of Colorado)	Oral 104 - Electrocatalytic Upcycling Of Nitrate Wastewater To Ammonia Fertilizer Via Electrified Membrane Jianan Gao (New Jersey Institute Of Technology)	Oral 110 - A Diafiltration Apparatus For High-Throughput Analysis Of Membrane Transport Properties Bill Phillip (University Of Notre Dame)	Oral 116 - Elucidating CO_2/Hydrocarbons Solubility Selectivity In Aliphatic Polycarbonates Via DFT Calculation And Experimentation Thien N Tran (University At Buffalo, The State University Of New York)
3:00 PM	Oral 93 - Peculiar Effect Of Low Loading Of Metal-Organic Polyhedra On CO_2/N_2 Separation Properties Of Cross-Linked Polyethers Taliehsadat Alebrahim (The State University Of New York At Buffalo)	Oral 99 - Purification Of Viral Vectors With Novel Membrane Adsorbers Jinxin Fan (North Carolina State University)	Oral 105 - Co-Transport Of Lithium, Sodium, And Potassium Ions In Polysulfone-Based Cems Jung Min Kim (University Of Virginia)	Oral 111 - New Insights In Ion Transport In Polyamide Membranes Using Impedance Spectroscopy Viatcheslav Freger (Technion - Israel Institute Of Technology)	Oral 117 - Non-Equilibrium Statistical Mechanics Approach For Extended Solution-Diffusion Model Albert Kim (University Of Hawai'i)
3:30 PM	Oral 94 - Cross-Linked Poly(Ionic Liquid)-Ionic Liquid Composite Membranes For CO_2/Light Gas Separations: Mixed-Gas Separation Performance And Long-Term Stability Under High-Temperature And -Pressure Conditions Chamaal Karunaweera (University Of Colorado Boulder)	Oral 100 - The Effects Of Low Flow And Flow Disruption On Virus Filtration Xianghong Qian (University Of Arkansas)	Oral 106 -Zwitterions And Their Influence Block Polymer Morphology Bradley Grim (Arizona State University)	Oral 112 - Interferometry For Precisely Measuring Ultralow Flow Rates From Permeable Materials Cody Ritt (Yale University)	Oral 118 - Mechanisms Of Ion And Water Transport In PEGDA Membranes Everett S Zofchak (The University Of Texas At Austin)
4:00 PM	Oral 95 - Directing Flexibility Of Metal-Organic Framework Toward Crystallographically Derived Molecular Sieving Ke Zhang (Aramco Americas)	Oral 101 - High Purification Of Binary Protein Mixtures Having Close Molecular Weights By Ultrafiltration Yufeng Song (New Jersey Institute Of Technology)	Oral 107 - Recent Progress In High Capacity, Stable, Lithium-Sulfur Batteries. Matthew R Hill (CSIRO And Monash University)	Oral 113 - The Influence Of Amine Structure And Water On The Mechanism Of CO_2 Facilitated Transport: Operando Characterization And Kinetic Modeling Casey O'Brien (University Of Notre Dame)	Oral 119 - Decoding The Molecular-Scale Determinants Of Antifouling At Polymer Membrane Selective Layers Dennis C Robinson Brown (The University Of California Santa Barbara)
4:30 PM	Oral 96 - Ethylene And Ethane Transport Properties Of Hydrogen-Stable Ag^+-Based Facilitated Transport Membranes Matthew N Davenport (The University Of Texas At Austin)	Oral 102 - Understanding The Mechanisms That Govern B. Diminuta Microfiltration To Improve The Sterile Filtration Of Therapeutic Viruses Evan Wright (Mcmaster University)	Oral 108 - Diffusion Coefficients Of Condensed Counterions In Ion-Exchange Membranes: Application Of Screening Length Scaling Relationship In Concentrated Electrolytes Yuxuan Huang (Columbia University)	Oral 114 - Inadequacy Of Current Approaches For Characterizing Membrane Transport Properties At High Salinities Yuanzhe Liang (Stanford University)	Oral 120 - Towards A Universal Framework For Evaluating Mass Transport In Pressure, Concentration, And Temperature Driven Membrane-Based Desalination Systems Kian P Lopez (University Of Colorado Boulder)

8:00 AM	Plenary Session III (Ballroom)				
Parallel Sessions	**Room Palm AD Award Session**	**Room Palm CF Membrane Fouling I**	**Room Abbey North Membrane Distillation and Pervaporation**	**Room Palm BE Industrial Applications**	**Room Abbey South Materials for Electrochemical Applications**
Chairs	**Caleb Funk** caleb.funk@dupont.com **Scott Husson** shusson@clemson.edu	**Steven T. Weinman** stweinman@eng.ua.edu **Ngoc T. Bui** ngoctbui21@ou.edu **Daniel J. Miller** danieljmiller@lbl.gov **Audie Thompson** akt022@uark.edu	**Lucy Mar. Camacho** Lucy.Camacho@tamuk.edu **Kailong Jin** Kailong.Jin@asu.edu	**Dibakar Bhattacharya** db@uky.edu **CJ Kurth** cj.kurth@gmail.com **Evan Hatakeyama** EHatakeyama@chevron.com	**William Tarpeh** wtarpeh@stanford.edu **Orlando Coronell** coronell@ad.unc.edu
9:30 AM	Oral 121 - Diglycolamide Membrane Adsorbers To Separate Lanthanides And Actinides For Use In Radiopharmaceuticals Christine E Duval (Case Western Reserve University)	Oral 127 - A Computational And Experimental Test Bed For Prediction Of RO Module Fouling Daniel J Miller (Lawrence Berkeley National Laboratory)	Oral 133 - Vapor Flux Of Membrane Distillation: Theoretical Limits, Insights On Membrane Design, And Anomalous Phenomena Ruoyu Wang (Vanderbilt University)	Oral 139 - An Enterprise Approach To Developing Industrial Membrane-Based Solutions Adil M Dhalla (Nanyang Technological University - Ntuitive Pte Ltd)	Oral 145 - Pressure Induced Diffusion In Hydrated Polymers Rahul Sujanani (The University Of Texas At Austin)
10:00 AM	Oral 122 - Putting Bubbles To Work: Next-Generation Water Treatment Systems Using Air-Trapping Membranes Anthony Straub (University Of Colorado Boulder)	Oral 128 - Application Of Machine Learning-Based Models To Understand And Predict Critical Flux Of Oil-In-Water Emulsion In Crossflow Microfiltration Henry J Tanudjaja (Nanyang Technological University)	Oral 134 - Modeling Pilot-Scale Membrane Distillation At High Salinity Using Bench-Scale Data Mukta Hardikar (The University Of Arizona)	Oral 140 - Single Pass TFF For Monoclonal Antibody Processing – Advancing A New Technology Through The MAST Center Andrew Zydney (Penn State University)	Oral 146 - Kinetic Control Of Intrinsic Pores In Monolayer Graphene For Large-Area Proton Selective Membranes Piran Kidambi (Vanderbilt University)
10:30 AM	Oral 123 - Elucidating The Role Of Micropore Generating Backbone Motifs And Amine Functionality On Sorption Energetics, Mixed-Gas Permeation, And H_2S Transport Katherine Mizrahi Rodriguez (Massachusetts Institute Of Technology)	Oral 129 - Quantifying Electrically Responsive Membranes For Biofouling Mitigation Charles-François De Lannoy (Mcmaster University)	Oral 135 - Integrated Electrocoagulation, Membrane Filtration And Membrane Distillation For Treatment Of Hydraulic Fracturing Produced Water Ranil Wickramasinghe (University Of Arkansas)	Oral 141 - Membranes Beyond Water - Deploying Spirals In Industrial Separations Benjamin Weaver (Solecta)	Oral 147 - Exploiting The Synergistic Effect Of Surfactants And Ionic Liquids For CO_2 Capture Reza Foudazi (University Of Oklahoma)
11:00 AM	Oral 124 - Selective Ion-Ion Separations And Desalination Using Adsorptive Ion-Capture Electrodialysis Membranes Adam A Uliana (University Of California, Berkeley)	Oral 130 - Feed Temperature Effects On Organic Fouling Of Reverse Osmosis Membranes: Competition Of Interfacial And Transport Properties Santiago Romero-Vargas Castrillon (University Of Edinburgh)	Oral 136 - Fermentation Broth Recovery Of Ethanol Using Polymer Based Membrane For Biofuel Production Oindrila Gupta (Oak Ridge National Laboratory)	Oral 142 - Separation And Recovery Of Critical Materials From End-Of-Life Lithium Ion Batteries Using Membrane Solvent Extraction Process Syed Islam (Oak Ridge National Laboratory)	Oral 148 - Computational Modeling Of 3D Electrospun Nanofiber Structures Albert Kim (University Of Hawai'i)
11:30 AM	Oral 125 - Engineering Subnanometer Vacancy Defects In Single-Layer Graphene For High-Performance Gas Separations Luis Francisco Villalobos (Yale University)	Oral 131 - Acoustically Excited Microstructure For On-Demand Fouling Mitigation In A Microfluidic Membrane Filtration Device Kieran R Fung (University Of Colorado Boulder)	Oral 137 - Impact Of Oxidative Chemicals On The Performance And Materials Properties Of Hydrophobic Porous Membranes Used In Membrane Distillation Elizabeth Hjelvik (University Of Colorado Boulder)	Oral 143 - Industrially Relevant Academic Research – Answering Fundamental Questions At The Module Scale Jeffrey R Mccutcheon (University Of Connecticut)	Oral 149 - Selective Separation Of Lithium Using Chemically Modified Nanoporous Polyamide Membrane With Electrodialysis Xi Chen (Stanford University)
12:00 PM	Oral 126 - Designing Cation-Exchange Membranes With Coordination-Based Selectivity Between Cations Ryan Duchanois (Yale University)	Oral 132 - Unraveling The Mechanism Of An AC Electric Field Applied For Fouling Mitigation Saikat Bhattacharjee (Technion-Israel Institute Of Technology)	Oral 138 - Porous Metal Foam Condensation Doubles Energy Efficiency Of Membrane Distillation Hamid Fattahijuybari (Purdue University)	Oral 144 - Facilitated Transport Membranes For Low Energy And Low Cost Carbon Capture And Sequestration Christine Parrish (Compact Membrane Systems)	Oral 150 - Counter-Ion Activation Energy Of Diffusion In Ion Exchange Membranes David Kitto (University Of Michigan)

Wednesday Afternoon

Parallel Sessions	Room Palm BE Seawater Desal and Osmotic Processes	Room Palm CF Membrane Fouling II	Room Palm AD Contaminant Removal from Water	Room Abbey North Water Reuse	Room Abbey South Module Modeling and Design
Chairs	**Andrea Achilli** achilli@email.arizona.edu **Milad Esfahani** mesfahani@eng.ua.edu **Jonathan Maisonneuve** maisonneuve@oakland.edu **William Phillip** wphillip@nd.edu	**Steven T. Weinman** stweinman@eng.ua.edu **Ngoc T. Bui** ngoctbui21@ou.edu **Daniel J. Miller** danieljmiller@lbl.gov **Audie Thompson** akt022@uark.edu	**Anthony Straub** tonystraub90@gmail.com **Andre da Costa**[i] adacosta@mtu.edu **Prakhar Prakash**[o] Prakhar.Prakash@chevron.com **Boya Xiong**[o] bxiong@umn.edu	**Vicky Karanikola** vkaranik@email.arizona.edu **Jack Gilron** jgilron@bgu.ac.il	**David Ladner** ladner@clemson.edu **Grigorios Panagakos** gpanagak@andrew.cmu.edu
2:00 PM	Oral 151 - Interfacial Polymerization Kinetics: New Insights On Film Formation Using In-Situ Microscopy And Particle-Tracking Adi Ben Zvi (Technion - Israel Institute Of Technology)	Oral 157 - Fouling-Resistant Membranes With Tunable Pore Size Fabricated Using Cross-Linkable Copolymers With High Zwitterion Content Samuel J Lounder (Tufts University)	Oral 163 - Development Of Highly Selective Biocatalytic Membrane Reactors For The Degradation Of Pesticides Lidietta Giorno (National Research Council Of Italy)	Oral 169 - Electroconductive Filters And Membranes: Future For Desalination And Wastewater Treatment Swatantra Pratap Singh (Indian Institute Of Technology Bombay)	Oral 175 - Computational Fluid Dynamics (CFD) Modeling Of 3D-Printed Spacers And Patterned Membranes In Reverse Osmosis And Nanofiltration Modules David Ladner (Clemson University)
2:30 PM	Oral 152 - A Solution-Diffusion With Defects Model To Explain Pressure-Dependent Membrane Permeability In Pressure Retarded Osmosis Zachary Binger (The University Of Arizona)	Oral 158 - Combined In-Situ Imaging And CFD To Unravel The Scaling Mechanism In Membrane Distillation Of High Salinity Brine Mahdi Malmali (Texas Tech University)	Oral 164 - Negative Rejection In Polyelectrolyte Multilayer Nanofiltration Membranes For Selective Ion Recovery Brielle Januszewski (Yale University)	Oral 170 - Chlorination And Autopsy Of Fouled Engineering-Scale Reverse Osmosis Membranes Bianca M. Souza Chaves (The University Of Arizona)	Oral 176 - Effect Of Packing Nonuniformity On The Performance Of Hollow Fiber Membrane Gas Separation Modules Fabricated From Fiber Tows Glenn Lipscomb (University Of Toledo)
3:00 PM	Oral 153 - Elucidating The Roles Of Polyamide Layer Structural Properties In The Permeability-Selectivity Tradeoff Governing Aqueous Separations Xi Chen (Stanford University)	Oral 159 - Fouling As An Opportunity: Exploiting The Properties Of Ultrathin Membranes To Make Fouling-Based Sensors James Mcgrath (University Of Rochester)	Oral 165 - Mxene-Functionalized Thin-Film Composite Membranes For PFAS Removal From Water Tin Le (The University Of Alabama)	Oral 171 - A Balanced-Charged Polyelectrolyte Complex Nanofiltration Membrane: Mitigating Scaling In Effluents Desalination Hao Huang (Ben Gurion University Of The Negev)	Oral 177 – Examining The Impact Of Acid Strength On Process Efficiency And Product Usability In Ammonia Recovery From Wastewater Using Membrane Contactors Abhishek Dutta (University Of British Columbia)
3:30 PM	Oral 154 - The Significance Of Membrane Charge And Co-Ion Partitioning In Determining Salt Transport Through Reverse Osmosis Membranes Li Wang (Yale University)	Oral 160 - Reactive Reverse Osmosis Membranes For Dissolved Silica Antifouling And Concentration Polarization Reduction By Catalytical Degradation Of Hydrogen Peroxide Weiliang Bai (University Of Texas At Austin)	Oral 166 - Graft Polymerization Using Oppositely Charged Monomers On NF Membrane For The Efficient Removal Of Carbamazepine From Wastewater Effluents Brhanu Kelali Desta (Ben-Gurion University Of The Negev)	Oral 172 - Degradation Of Polyamide Thin Film By Chlorine And Peracetic Acid: A Combined QCM-D And AFM Study Tashfia M Mohona (University At Buffalo)	Oral 178 – Accelerating The Optimization Of Turbulence-Promoting Spacers Using Computational Fluid Dynamics, Surrogate Modeling, And Machine Learning Zachary Binger (The University Of Arizona)
4:00 PM	Oral 155 - Forward Osmosis And Freeze Concentration For Mining And Metals Effluents Noel Devaere (University Of Toronto)	Oral 161 - Scalable Antifouling Membranes Via Zwitterionic Polymer Brushes: Examining The Roles Of Brush Thickness And Density Allyson L Mcgaughey (Princeton University)	Oral 167 - Modeling Of RO System Water Treatment Operation For Nitrate And Salt Removal Using Long-Short Term Memory (LSTM) Machine Learning Model With Attention Coefficient Yoram Cohen (University Of California, Los Angeles)	Oral 173 - Solute-Solute Selectivity In Membrane Separation: Definition, Measurement, And Process Scale Evaluation Ruoyu Wang (Vanderbilt University)	Oral 179 – Flow Dynamics In A Spacer-Filled Membrane Distillation Channel Mahdi Malmali (Texas Tech University)
4:30 PM	Oral 156 - Predicting Efficiency And Elucidating Mechanisms For Organic And Inorganic Compound Removal By Reverse Osmosis And Nanofiltration Membranes Using Machine Learning Nohyeong Jeong (Colorado State University)	Oral 162 - Raman Chemical Fingerprints For Real-Time Detection Of Membrane Fouling Danielle Park (University Of Colorado Boulder)	Oral 168 - Nanofiltration And Temperature Swing Reverses Osmosis For The Removal Of Herbicide 2-Methyl-4-Chlorophenoxyacetic Acid (MCPA) From Saline Industrial Wastewater Mikel Duke (Victoria University, Melbourne)	Oral 174 - Effect Of Pore Defects On Membrane Rejection In Pilot-Scale Membrane Distillation Mukta Hardikar (The University Of Arizona)	Oral 180 – Organic Matter Removal For Reverse Osmosis Fouling Mitigation: From Lab Prototype And Numerical Simulation To Pilot Plant Yunqiao Ma (Nanyang Technological University)

* o - online i - in absentia

Ballroom, 8:00 am - 9:00 am, Monday, May 16th

Separation Membranes Incorporating Metal–Organic Frameworks and Porous Network Polymers

Plenary Speaker

Dr. Jeffrey R. Long

Chemistry and Chemical and Biomolecular Engineering, University of California, Berkeley

Jeffrey R. Long is a Professor of Chemistry and Chemical & Biomolecular Engineering at the University of California, Berkeley and a Faculty Senior Scientist in the Materials Sciences Division at Lawrence Berkeley National Laboratory. He served as Chair of the Division of Inorganic Chemistry of the American Chemical Society in 2012 and as a founding Associate Editor of the journal Chemical Science. He co-founded and directs two companies: Mosaic Materials, which is developing metal–organic frameworks for low-energy carbon dioxide capture, and Flux Technology, which is producing high-performance polymer membranes for gas purification. His 375 publications have received more than 83,000 citations, and his recent awards include election to the American Academy of Arts and Sciences, the 2019 American Chemical Society F. Albert Cotton Award in Synthetic Inorganic Chemistry, and the 2020 Royal Society of Chemistry Ludwig Mond Award.

Abstract

The extraordinary chemical tunability and high surface areas of metal–organic frameworks (MOFs) has enabled realization of powerful new materials for adsorptive gas separations. For example, materials of the type M_2(dobdc) (M = Mg, Mn, Fe, Co, Ni Cu, Zn; dobdc = 2,5-dioxido-1,4-benzenedicarboxylate) possess a record high density of coordinatively-unsaturated metal sites that can facilitate olefin/paraffin separations at near ambient temperatures. Incorporation of nanocrystals of these materials into polyimide membranes enhances selectivity for ethylene over ethane, while also increasing ethylene permeability owing to the high porosity. Similar enhancements are apparent in related membranes studied for potential use in the removal of CO_2 from natural gas. Remarkably, the strong interaction between the MOF nanocrystals and the polymer improves membrane stability to plasticization, permitting selectivity to be maintained at higher feed pressures. More recently, we have investigated the effects of incorporating particles of porous network polymers, specifically the porous aromatic framework PAF-1, as a general means of significantly increasing permeability in a polymer membrane while maintaining selectivity for a wide range of gas separations. Given its chemical stability and the possibility of functionalizing its pore surfaces with a high density of selective binding groups, PAF-1 further provides a platform for the targeted removal of a variety of species from aqueous solution. Indeed, derivatives that can take up lanthanides, copper, iron, mercury, or boron rapidly, at high capacity, and with high selectivity have now been demonstrated. Additionally, we have shown that such PAF particles can be incorporated into ion exchange polymers to create membranes capable of targeted impurity capture while desalinating various water sources via electrodialysis.

Ballroom, 8:00 am - 9:00 am, Tuesday, May 17th

Membrane à la carte

Plenary Speaker

Dr. Kitty Nijmeijer

Chemical Engineering and Chemistry, Eindhoven University of Technology

Prof. Dr. Ir. Kitty Nijmeijer is professor of Membrane Materials and Processes at Eindhoven University of Technology. She received her PhD from the University of Twente (NL) where she became professor Membrane Science and Technology in 2015. In 2016, together with her colleague Dr. Zandrie Borneman, she got the opportunity to build and serve as the chair of Membrane Materials and Processes at Eindhoven University of Technology (NL), which now has ~20 researchers, 3 scientific staff and 3 supportive personnel, and a fully equipped, state-of-the-art membrane lab (~300 m^2). She is dedicated to molecular design and the development of polymer membranes to separate molecular mixtures in sustainable processes. Her research merges materials chemistry, membrane design and formation, and membrane process technology. Since 2010 she published over 120 papers that are categorized in three application areas: 1) Water purification and water treatment; 2) Energy generation and storage; 3) Energy efficient gas separation. Moreover, she was a member of the board of reviewing editors of the scientific journal Science, and an editorial board member of Journal of Membrane Science. She chaired the International Conference on Membranes (ICOM 2011; ~1200 participants), the leading scientific conference on membranes. She serves as the vice-dean of the Department of Chemical Engineering and Chemistry at the Eindhoven University of Technology since 2018. She is also the winner of the 2018 Academic Society Award of the Dutch Royal Institute of Engineers (KIVI) and recorder two TEDx talks and numerous interviews on the national television to share her passion for science and technology.

Abstract

With the transition to a sustainable, circular economy and a strong emphasis on the recovery and reuse of materials, minerals and nutrients, membrane technology will become increasingly important. Associated to this, the complexity of separations will increase requiring membranes with molecular selectivity suitable for the specific recovery of target valuable components. This comes with demanding membrane characteristics in terms of properties and separation performance. Obviously with this in mind, dedicated membrane design and tunable membrane development and production methods are essential.

In this presentation, several methods to tailor membrane morphology and separation properties in the framework of recovery and reuse are discussed. Polymer membranes are currently often prepared using methods like phase separation and interfacial polymerization. Although phase separation was already extensively explored about a decade ago, this method still offers multiple directions to tune membrane morphology. This allows to a certain extent control over pore size and membrane polymer-component interactions, but real molecular separations are out of reach still as phase separation results in broad pore size distributions and interactions are limited due to the use of only few membrane polymer chemistries. To move beyond these limitations, several new concepts explored recently for the preparation of membranes are evaluated, starting from methods to tailor retention and permeability at a component level, methods to produce highly porous membrane structures or bi-continuous polymer films are presented. Finally, the role of self-organizing materials as a route to true molecular selectivity is discussed. Parallel to this, multiple examples of the characteristics and performance of membranes developed with these advancing methods are given.

Ballroom, 8:00 am - 9:00 am, Wednesday, May 18th

Advanced Membrane Technologies for Clean Energy Transition: Innovations and Challenges

Plenary Speaker

Dr. Chunqing Liu

Sr. Manager of Green H2 Growth Vector Group, Honeywell UOP.

Dr. Liu is the R&D Fellow and Sr. Manager of Green Hydrogen Growth Vector Group at Honeywell UOP. She leads a multi-disciplinary research group in the fields of membrane technology, electrochemistry, catalysis, and materials science for the development of breakthrough water electrolysis technologies to produce green hydrogen. Prior to her current appointment, she led the Membranes R&D Group at UOP to develop novel membranes to advance gas separations, energy storage, and green hydrogen sustainable technologies. She has published 5 book chapters, over 60 peer-reviewed papers, and holds over 100 patents. She received the Honeywell Excellence in Innovation Award in 2010, the ACS Women Chemists Committee Rising Star Award in 2015, Honeywell UOP Stine Star Award in 2019, and Honeywell Innovation Initiators Award in 2022. Dr. Liu received her PhD in Polymer Chemistry and Physics from the Institute of Chemistry, the Chinese Academy of Sciences in Beijing, China, and worked as a postdoctoral researcher at Northwestern University and the University of Illinois at Urbana-Champaign before she joined Honeywell UOP.

Abstract

Clean energy transition to achieve net zero global emissions by midcentury requires new technologies and investments. Innovation has been the engine for the development of breakthrough membrane technologies for clean energy transition at Honeywell UOP. Advances in membrane materials, fabrication methods, module design, process design, and integration of membrane with other technologies have been crucial to the increased industrial adoption of UOP's SeparexTM membrane technology for gas separations. This talk will review the design and development of several generations of UOP SeparexTM commercial membranes for large-scale natural gas upgrading, enhanced oil recovery, and hydrogen purification and recovery applications. Our recent efforts for the development of new hollow fiber membrane for biogas purification, novel membranes and membrane-electrode assembly (MEA) for flow battery application for energy storage, as well as new catalyst-coated membranes (CCMs) for water electrolysis for the production of green hydrogen will also be discussed.

Oral Sessions - Monday

Presenting Authors Are Indicated By An Asterisk*. See The Most Up-To-Date Version Of The Program Book At **www.Membranes.Org/Nams~2022/**

Oral 1 – Organic Solvent Separations I

9:30 Am – 12:30 Pm, Room Palm AD

Session Chairs

Michele Galizia
The University Of Oklahoma

Steve White
Membrane Technology & Research

Neel Rangnekar
Exxonmobil

9:30 Am (Palm AD, Monday)

Oral 1 - From Renewable Building Blocks To Biodegradable Polymer Membranes With Solvent-Resistant Properties (In-Person)

Gyorgy Szekely (King Abdullah University Of Science And Technology)*

10:00 Am (Palm AD, Monday)

Oral 2 - Towards Wired-Lab Organic Solvent Nanofiltration: Is Machine Learning Viable? (Virtually)

Gergo Ignacz (King Abdullah University Of Science And Technology)*; Gyorgy Szekely (King Abdullah University Of Science And Technology)

10:30 Am (Palm AD, Monday)

Oral 3 - Tuning Polyimide Thin Film Composite Membranes For Organic Solvent Reverse Osmosis Separations Via Boc Protected Amine Solid-State Crosslinking (In-Person)

Yacine Feliachi (Georgia Tech)*; M.G. Finn (Georgia Institute Of Technology); Ryan Lively (Georgia Tech)

11:00 Am (Palm AD, Monday)

Oral 4 - Organic Solvent Stable Graft Modified Polybenzimidazole Membranes With Customisable Rejection Profiles (In-Person)

Adam E Oxley (Imperial College London)*; Piers Gaffney (Queen Mary University Of London); Andrew Livingston (Queen Mary University Of London)

11:30 Am (Palm AD, Monday)

Oral 5 - Precision Molecular Sieving With Mof And Cof Membranes In Organic Solvents (Virtually)

Rifan Hardian (King Abdullah University Of Science And Technology)*; Gyorgy Szekely (King Abdullah University Of Science And Technology)

12:00 Pm (Palm AD, Monday)

Oral 6 - Engineering Organic Solvent Reverse Osmosis Performance Of Vapor Phase Infiltrated Organic-Inorganic Hybrid Membranes (In-Person)

Yi Ren (Georgia Institute Of Technology)*; Emily Mcguinness (Georgia Institute Of Technology); Benjamin Jean (Georgia Institute Of Technology); Chaofan Huang (Georgia Institute Of Technology); Roshan Joseph (Georgia Institute Of Technology); Mark Losego (Georgia Institute Of Technology); Ryan Lively (Georgia Tech)

Oral 2 – Carbon Capture I

9:30 Am – 12:30 Pm, Room Palm CF

Session Chairs

Winston Ho
Ohio State University

David Hopkinson
National Energy Technology Laboratory

Katherine Hornbostel
University Of Pittsburg

Chen Zhang
University Of Maryland

9:30 Am (Palm CF, Monday)

Oral 7 - Thin Film Composite Membranes With Novel Quaternary Ammonium Functionalized Poly(Arylene Ether Sulfone) Copolymers For Direct Air Capture (In-Person)

Hoda Shokrollahzadeh Behbahani (Arizona State University)*; Matthew Green (Arizona State University)

10:00 Am (Palm CF, Monday)

Oral 8 - Integrated Facilitated Transport Membrane Modules For Highly Selective Syngas Purification And Carbon Capture (In-Person)

Yang Han (The Ohio State University)*; Yutong Yang (The Ohio State University); Ruizhi Pang (The Ohio State University); Winston Ho (Osu)

10:30 Am (Palm CF, Monday)

Oral 9 - Thin-Film Composite Membranes Based On Hyperbranched Poly(Ethylene Oxide) For CO_2/N_2 Separation (In-Person)

Gengyi Zhang (The State University Of New York At Buffalo)*; Thien N Tran (University At Buffalo, The State University Of New York); Liang Huang (The State University Of New York At Buffalo); Erda Deng (State University Of New York At Buffalo); Adrienne Blevins (Materials Science And Engineering Program, University Of Colorado Boulder); Wenji Guo (University At Buffalo, The State University Of New York); Yifu Ding (Membrane Science, Engineering, And Technology (Mast) Center, Department Of Mechanical Engineering, University Of Colorado Boulder); Haiqing Lin (The State University Of New York At Buffalo)

11:00 Am (Palm CF, Monday)

Oral 10 - Polyvinylpyridine-Based Catalytic Membranes For Integrated CO_2 Capture And Conversion: Structure-Function Relationships (In-Person)

Casey O'brien (University Of Notre Dame)*

11:30 Am (Palm CF, Monday)

Oral 11 - Effect Of Free Volume Modification (Fvm) On The Physical Aging Behavior Of Amine-Functionalized Pim-1 Membranes For Gas Separation (In-Person)

Taigyu Joo (Massachusetts Institute Of Technology)*; Katherine Mizrahi Rodriguez (MIT); Hyunhee Lee (MIT); Zachary Smith (MIT)

12:00 Pm (Palm CF, Monday)

Oral 12 - The Effect Of Channel Height Variation On CO_2 Separation Performance In A Flat Sheet Membrane (In-Person)

Joanna Rivero (University Of Pittsburgh)*; Marcos Da Conceicao (University Of Toledo); Leo Nemetz (University Of Toledo); Glenn Lipscomb (University Of Toledo, Toledo); Katherine Hornbostel (University Of Pittsburgh)

Oral 3 – Emerging Materials For Liquid Separation I

9:30 Am – 12:30 Pm, Room Palm BE

Session Chairs

Oishi Sanyal
West Virginia University

Baoxia Mi
University Of Berkeley, California

Beza Getachew
Rice University

9:30 Am (Palm BE, Monday)

Oral 13 - Separation Of Trace Organics From Aqueous Streams Using Carbonaceous Membranes (In-Person)

Haley D White (Georgia Institute Of Technology)*; Ryan Lively (Georgia Tech)

10:00 Am (Palm BE, Monday)

Oral 14 - Cucurbit[N]Uril-Based Nanofiltration Membranes For Effective Molecular Separation (Virtually)

Shi-Peng Sun (Nanjing Tech University)*

10:30 Am (Palm BE, Monday)

Oral 15 - Nanoporous Multilayer Graphene Membrane For Precise And Stable Forward Osmosis Desalination (In-Person)

Claudio A Ruiz Torres (The University of Toronto)*; Jay Werber (The University Of Toronto)

11:00 Am (Palm BE, Monday)

Oral 16 - High Flux Vapor-Gap Membranes For Reverse Osmosis Via Informed Design Of Membrane Pore Structures (In-Person)

Elizabeth A Hjelvik (University Of Colorado Boulder)*; Kian P Lopez (University Of Colorado Boulder); Anthony Straub (University Of Colorado Boulder)

11:30 Am (Palm BE, Monday)

Oral 17 - Preparation Of Functionalized Porous Poly (Ether Ether Ketone), Peek, Materials And Hollow Fiber Membranes (In-Person)

Andrii Bazylevich (Ariel University); Sagiv Weintraub (Ariel University); Gary Gellerman (Ariel University); Ben Bikson (Avanpore)*

12:00 Pm (Palm BE, Monday)

Oral 18 - Polyethyleneimine-Based Surface Treatment For Facile Synthesis Of Conductive Membranes: Active Material Doesn't Matter (In-Person)

Wei Zhang (University Of North Carolina At Chapel Hill)*; Orlando Coronell (University Of North Carolina At Chapel Hill)

Oral 4 – Membrane Synthesis And Casting I

9:30 Am – 12:30 Pm, Room Abbey North

Session Chairs

Rachel Dorin
Terapore Technologies, Inc

Christine E. Duval
Case Western Reserve University

Abhishek Roy
National Renewable Energy Laboratory

Reza Foudazi
The University Of Oklahoma

9:30 Am (Abbey North , Monday)

Oral 19 - Microstructure Evolution During Nonsolvent-Induced Phase Separation: A Joint Experimental-Computational Investigation Into Membrane Formation (In-Person)

Alexander Bridge (The University Of Texas At Austin)*; Anthony J Cooper (University Of California, Santa Barbara); Jan Garcia (Uc Santa Barbara); Matthew Santoso (The University Of Texas At Austin); Glenn Fredrickson (Uc Santa Barbara); Benny Freeman (The University Of Texas At Austin)

10:00 Am (Abbey North , Monday)

Oral 20 - Automated Stepwise Synthesis Of Poly(Acrylonitrile)-Supported Oligoamide Membranes And Their Dye–Salt Separation (In-Person)

Paramita Manna (Ben-Gurion University Of The Negev); Roy Bernstein (Ben Gurion University); Ron Kasher (Ben Gurion University Of The Negev)*

10:30 Am (Abbey North , Monday)

Oral 21 - Microporous Polymer Membranes: A New Approach To Simultaneously Reducing Physical Aging And Swelling While Enhancing Permeability (In-Person)

Jing Deng (University Of Oklahoma); Cara Doherty (Csiro); Laura Matesanz Nino (University Of Valladolid); Angel Lozano (University Of Valladolid); Cristina Alvarez (University Of Valladolid); Anita J. Hill (Csiro); Michele Galizia (University Of Oklahoma)*

11:00 Am (Abbey North , Monday)

Oral 22 - Crosslinked Triptycene-Containing Polybenzoxazole Membranes With Model Network Structures For High Performance Gas Separations (In-Person)

Ziwei Dai (University Of Notre Dame)*

11:30 Am (Abbey North , Monday)

Oral 23 - Integrated Roll-To-Roll (R2R) Fabrication Of Graphene Oxide Nanofiltration Membranes For Applications In Extreme Operating Environments (In-Person)

Amirsalar Rabbani Esfahani (Georgia Institute Of Technology)*; Chen Ma (Georgia Institute Of Technology); Uwezo Flewellen (Georgia Institute Of Technology); Sankar Nair (Georgia Institute Of Technology); Tequila Harris (Georgia Institute Of Technology)

12:00 Pm (Abbey North , Monday)

Oral 24 - Super-Hydrophobic MFI Zeolite Hollow Fiber Membranes For Alcohol/Water Separation (In-Person)

Seyed Mirfendereski (Arizona State University); Jerry Lin (Arizona State University)*

Oral 5 – Inorganic Materials

9:30 Am – 12:30 Pm, Room Abbey South

Session Chairs

Luis Francisco Villalobos
Yale University

Jay Kniep
Membrane Technology & Research

Minghui Qiu
Nanjing Tech University

David Bergsman
University Of Washington

9:30 Am (Abbey South , Monday)

Oral 25 - High-Pressure CO_2 Permeation Properties Of Ceramic-Carbonate Dual-Phase Membranes (In-Person)

Oscar Ovalle (Arizona State University); Jerry Lin (Arizona State University)*

10:00 Am (Abbey South , Monday)

Oral 26 - Metal Ions-Doped Ultrahigh Flux Composite Graphene Oxide/Carbon Nanotubes (Go-Swcnt) Membranes For Efficient H_2 Separation (In-Person)

Daniel Gardeno (University Of Chemistry And Technology Prague); Zdeněk Sofer (University Of Chemistry And Technology Prague); Daniel Bouša (University Of Chemistry And Technology Prague); Saeed Jamali Ashtiani (University Of Chemistry And Technology Prague); Karel Friess (University Of Chemistry And Technology Prague)*

10:30 Am (Abbey South , Monday)

Oral 27 - Network Nanostructured Zif-8 To Enable Percolation And Enhanced Gas Transport (In-Person)

Hyunhee Lee (MIT)*; Won Seok Chi (Chonnam National University); Moonjoo Lee (Samsung); Ke Zhang (Aramco Americas); Katherine Mizrahi Rodriguez (MIT); Stephen Dewitt (MIT); Zachary Smith (MIT)

11:00 Am (Abbey South , Monday)

Oral 28 - High-Performance CO_2- And O_2-Sieving By Controlled Etching Of Graphene (In-Person)

Shiqi Huang (Ecole Polytechnique Fédérale De Lausanne)*; Shaoxian Li (Epfl); Luis Francisco Villalobos (Yale University); Kumar Varoon Agrawal (Ecole Polytechnique Fédérale De Lausanne)

11:30 Am (Abbey South , Monday)

Oral 29 - Tunable Organosilica Membranes For High-Temperature H_2/CO_2 Separation Synthesized By Rapid Oxygen Plasma Treatment Of Polysiloxane (In-Person)

Vinh Bui (University At Buffalo)*; Lingxiang Zhu (U.S. Department Of Energy National Energy And Technology Laboratory); Leiqing Hu (University At Buffalo); Liang Huang (The State University Of New York At Buffalo); Elizabeth Haddad (University At Buffalo); Erda Deng (State University Of New York At Buffalo); Thien N Tran (University At Buffalo, The State University Of New York); Haiqing Lin (The State University Of New York At Buffalo)

12:00 Pm (Abbey South , Monday)

Oral 30 - Synthesis Of Zif Membranes By A Vapor-Phase Seeding Method For Olefin/Paraffin Separation (In-Person)

Zhiqin Qiang (University Of Wisconsin-Milwaukee)*; Weiling Xia (University Of Wisconsin-Milwaukee); Rachel Clark (University Of Wisconsin-Milwaukee); Zihao Yi (University Of Wisconsin-Milwaukee); Rahul Sampat Khandge (University Of Wisconsin-Milwaukee); Xiaoli Ma (University Of Wisconsin-Milwaukee)

Oral Sessions - Monday

Presenting authors are indicated by an asterisk*. See the most up-to-date version of the program book at **www.membranes.org/nams~2022/**

Oral 6 – Organic Solvent Separations II

2:00 pm – 5:00 pm, ROOM Palm AD

Session Chairs

Michele Galizia
The University of Oklahoma

Steve White
Membrane Technology & Research

Neel Rangnekar
ExxonMobil

2:00 pm (Palm AD, Monday)

Oral 31 - Interfacial Polymerization With Hydrophobic Amines Provides Ultrathin Polyamide Nanofilms For Hydrocarbon Separation (In-Person)

Neel Rangnekar (Exxonmobil)*; Siyao Li (Imperial College London); Ruijiao Dong (Imperial College London); Valentina Elena Musteata (King Abdullah University Of Science And Technology (KAUST)); Jihoon Kim (Queen Mary University Of London); JR Johnson (Exxon Mobil); Bennett Marshall (Exxonmobil); Stefan Chisca (KAUST); Jia Xu (Imperial College London); Scott Hoy (Exxonmobil Research And Engineering); Benjamin Mccool (Exxonmobil Research And Engineering); Suzana Nunes (King Abdullah University Of Science And Technology (KAUST)); Zhiwei Jiang (Queen Mary University Of London); Andrew Livingston (Queen Mary University Of London)

2:30 pm (Palm AD, Monday)

Oral 32 - A Smart And Responsive Crystalline Porous Organic Cage Membrane With Switchable Pore Apertures For Graded Molecular Sieving (Virtually)

Zhiwei Jiang (Queen Mary University Of London)*; Ai He (University Of Liverpool); Hadeel Hussain (Diamond Light Source); Jonathan Rawle (Diamond Light Source); Michael Briggs (University Of Liverpool); Yue Wu (University Of Liverpool); Marc Little (University Of Liverpool); Andrew Livingston (Queen Mary University Of London); Andrew Cooper (University Of Liverpool)

3:00 pm (Palm AD, Monday)

Oral 33 - Solution Processable Polytriazoles From Spirocyclic Monomers For Membrane-Based Hydrocarbon Separations (In-Person)

Nicholas C Bruno (Georgia Institute Of Technology)*; M.G. Finn (Georgia Institute Of Technology); Ryan Lively (Georgia Tech); Ronita Mathias (Georgia Institute Of Technology)

3:30 pm (Palm AD, Monday)

Oral 34 - Organometallic Hybrid Nanofilms With Highly Interconnected Pore Architecture For Ultrafast Solvent Transport With Precise Molecular Separation (In-Person)

Bratin Sengupta (University At Buffalo)*; Qiaobei Dong (Gas Technology Institute); Ji Jiang (Rensselaer Polytechnic Institute); Dinesh Behera (University At Buffalo); Fanglei Zhou (Membrane Technology And Research); Miao Yu (University At Buffalo)

4:00 pm (Palm AD, Monday)

Oral 35 - Effect Of Xylene Activity And Crystal Orientation Of MFI- Zeolite Membranes On Separation Of Xylene Isomers (In-Person)

Fateme Banihashemi (Arizona State University); Jerry Lin (Arizona State University)*

4:30 pm (Palm AD, Monday)

Oral 36 - Data-Driven Development Of Polymer Membrane Materials For Complex Mixture Separations (In-Person)

Youngjoo Lee (Georgia Tech)*; Lihua Chen (Georgia Tech); Rampi Ramprasad (Georgia Tech); Ryan Lively (Georgia Tech)

Oral 7 – CARBON CAPTURE II

2:00 pm – 5:00 pm, ROOM Palm CF

Session Chairs
Winston Ho
Ohio State University

David Hopkinson
National Energy Technology Laboratory

Katherine Hornbostel
University of Pittsburg

Chen Zhang
University of Maryland

2:00 pm (Palm CF, Monday)

Oral 37 - Direct Ocean Capture Using Membrane Contactors (In-Person)

Austin R Lieber (University Of Pittsburgh)*; Joanna Rivero (University Of Pittsburgh); Donna Hildebrandt (University Of Pittsburgh); Chris Snodgrass (University Of Pittsburgh); William Gamble (University Of Pittsburgh); Zoe Neal (University Of Pittsburgh); Shanna Davidson (University Of Pittsburgh); Huda Usman (University Of Pittsburgh); Tagbo Niepa (University Of Pittsburgh); Katherine Hornbostel (University Of Pittsburgh)

2:30 pm (Palm CF, Monday)

Oral 38 - Highly Permeable Substrates Prepared By Vapor-Induced Phase Separation For CO_2-Selective Composite Membranes (In-Person)

Ruizhi Pang (The Ohio State University)*; Yutong Yang (The Ohio State University); Yang Han (The Ohio State University); Kai Chen (The Ohio State University); Winston Ho (OSU)

3:00 pm (Palm CF, Monday)

Oral 39 - Techno-Economic Assessment Of Postcombustion Capture Processes Using High-Performance Nanoporous Graphene Membranes (Virtually)

Marina Micari ((Ecole Polytechnique Fédérale De Lausanne)*; Kumar Varoon Agrawal (Ecole Polytechnique Fédérale De Lausanne)

3:30 pm (Palm CF, Monday)

Oral 40 - Critical Aspects Of High-Pressure CO_2-Induced Plasticization In Polyimide Membranes (In-Person)

Menno Houben (Eindhoven University Of Technology)*; Romy Van Geijn (Eindhoven University Of Technology); Joey Kloos (Eindhoven University Of Technology); Machiel Van Essen (Eindhoven University Of Technology); Zandrie Borneman (Eindhoven University Of Technology); Kitty Nijmeijer (Eijndhoven)

4:00 pm (Palm CF, Monday)

Oral 41 - In Situ Synergistic Growth Of Crystalline And Polymer-Incorporated Amorphous ZIF-8 In Polybenzimidazole Achieving Hierarchical Nanostructures For H_2/CO_2 Separation (In-Person)

Leiqing Hu (University At Buffalo)*; Vinh Bui (University At Buffalo); Haiqing Lin (The State University Of New York At Buffalo)

4:30 pm (Palm CF, Monday)

Oral 42 - Mitigated Carrier Saturation Of Facilitated Transport Membranes For Decarbonizing Dilute CO_2 Sources (In-Person)

Yang Han (The Ohio State University)*; Winston Ho (The Ohio State University)

Oral 8 – EMERGING MATERIALS FOR LIQUID SEPARATION II

2:00 pm – 5:00 pm, ROOM Palm BE

Session Chairs
Oishi Sanyal
West Virginia University

Baoxia Mi
University of Berkeley, California

Beza Getachew
Rice University

2:00 pm (Palm BE, Monday)

Oral 43 - Bioinspired Cross-Linked Iron Doped Reduced Graphene Oxide (rGO) Membrane For Ultrafiltration Of Organic Micropollutants (Virtually)

Yichen Wu (Mcmaster University); Nan Zhang (Mcmaster University); Gavin Yuen (Mcmaster University); Charles-François De Lannoy (Mcmaster University)*

2:30 pm (Palm BE, Monday)

Oral 44 - Functional Nanofibrous Membranes Modified With Β-Cyclodextrins For Improved Surface Characteristics And Filtration Performance (Virtually)

Saranya Rameshkumar (Trinity College Dublin)*; Ramesh Babu Padamati (Trinity College Dublin)

3:00 pm (Palm BE, Monday)

Oral 45 - Cross-Linkable/ Self-Assembling Terpolymer Membranes For Brackish Water Desalination (In-Person)

Samuel J Lounder (Tufts)*; Ayse Asatekin (Tufts)

3:30 pm (Palm BE, Monday)

Oral 46 - Including Non-Binding 'Spacer' Monomers In Polyprotic Polymeric Ligands Impacts Ligand-Ion Affinity For Lanthanum (In-Person)

Priyanka Suresh (Case Western Reserve University)*; Christine E Duval (Case Western Reserve University); Ming Yu (Case Western Reserve University); Amy Che (Case Western Reserve University); Kevin Pataroque (Case Western Reserve University); Dylan Kulbacki (Case Western Reserve University)

4:00 pm (Palm BE, Monday)

Oral 47 - Enhancing The Salt Rejections Of Covalent Organic Framework Membranes By Creating Dual-Layer Membrane Structure (In-Person)

Miguel Jaimes (University Of Wisconsin-Milwaukee)*; Zhiqin Qiang (University Of Wisconsin-Milwaukee); Rahul Khandge (University Of Wisconsin - Milwaukee); Ho Kuan Yu (University Of Wisconsin - Milwaukee); Weiling Xia (University Of Wisconsin - Milwaukee); Rachel Clark (UW - Milwaukee); Xiaoli Ma (University Of Wisconsin-Milwaukee)

4:30 pm (Palm BE, Monday)

Oral 48 - Polyampholyte Copolymer Self-Assembly For Fouling Resistant And Easily Tunable Membranes (In-Person)

Luca Mazzaferro (Tufts University)*; Samuel J Lounder (Tufts); Ayse Asatekin (Tufts)

Oral 9 – MEMBRANE SYNTHESIS AND CASTING II

2:00 pm – 5:00 pm, ROOM Abbey North

Session Chairs

Rachel Dorin
TeraPore Technologies, Inc

Christine Duval
Case Western Reserve University

Abhishek Roy
National Renewable Energy Laboratory

Reza Foudazi
The University of Oklahoma

2:00 pm (Abbey North, Monday)

Oral 49 - Thin Zeolitic-Imidazolate Framework ZIF-8 Membranes Supported On ZnO-Deposited Substrates For Propylene/Propane Separation (In-Person)

Anil Ronte (Oklahoma State University)*; Phadindra Wagle (Oklahoma State University); Diako Mahmodi (Oklahoma State University); SHAILESH DANGWAL (Oklahoma State University); Elena Echeverria (Oklahoma State University); Dave Mcilroy (Oklahoma State University); Seok Jhin Kim (Oklahoma State University)

2:30 pm (Abbey North, Monday)

Oral 50 - 3D Printing Of Thin Film Composite Membranes For Nanofiltration And Beyond (In-Person)

Jeffrey R Mccutcheon (University Of Connecticut)*; Xin Qian (University Of Connecticut); Mayur Ostwal (University Of Connecticut); Edward Wazer (University Of Connecticut); Noah Ferguson (University Of Connecticut); Marianne Pemberton (University Of Connecticut)

3:00 pm (Abbey North, Monday)

Oral 51 - Advanced Ionic Polymers For High-Performance Gas Separation Membranes (In-Person)

Sudhir Ravula (University Of Alabama)*; Jason Bara (University Of Alabama)

3:30 pm (Abbey North, Monday)

Oral 52 - Confined Thermally Induced Phase Separation under Micromolding – Fabrication of Patterned Membranes with Enhanced Pore Connectivity and High Salt-scaling Resistance (In-person)

Shouhong Fan (University Of Colorado At Boulder)*; Yifu Ding (University Of Colorado At Boulder); Anthony Straub (University Of Colorado At Boulder); Kamalesh K Sirkar (New Jersey Institute Of Technology); Adrienne Blevins (University Of Colorado At Boulder); Duong T. Nguyen (University Of Colorado At Boulder); Jaylene Martinez (University Of Colorado At Boulder); John Chau (New Jersey Institute Of Technology)

4:00 pm (Abbey North, Monday)

Oral 53 - Sustainability In Membrane Separations: Integration Of Eco-Friendly Materials And Scaled-Up Fabrication Of Polymeric Membranes For Water Purification (In-Person)

David Lu (University Of Kentucky)*; Kwangjun Jung (Georgia Institute Of Technology); Ju Yong Shim (Georgia Institute Of Technology); Alex Williams (University Of Kentucky); Kassandra Opfar (University Of Kentucky); Isabel Escobar (University Of Kentucky); Tequila Harris (Georgia Institute Of Technology)

4:30 pm (Abbey North, Monday)

Oral 54 - Polyamide Thickness And Morphology Features Of Thin-Film Composite Membranes Depend On The Extent Of Restriction Of The Amine Monomer Supply During Interfacial Polymerization (Virtually)

Mikayla D Armstrong (University Of North Carolina At Chapel Hill)*; Kasia Grzebyk (UNC Chapel Hill); Orlando Coronell (University Of North Carolina At Chapel Hill)

Oral 10 – Catalytic and Responsive Membranes

2:00 pm – 5:00 pm, ROOM Abbey South

Session Chairs

Miao Yu
University at Buffalo

Ayse Asatekin
Tufts Univeristy

Simona Liguori
Clarkson University

Volodymyr Tarabara
Michigan State University

2:00 pm (Abbey South, Monday)

Oral 55 - Synthesis Of Reactive Membranes For Capture And Degradation Of Water And Air Pollutants (In-Person)

Rollie G Mills (University Of Kentucky)*; Dibakar Bhattacharyya (U. Kentucky); Matthew D Bernard (University Of Kentucky); Ronald J. Vogler (University Of Kentucky); Jacob Concolino (University Of Kentucky); Todd Hastings (University Of Kentucky); Yinai Wei (University Of Kentucky); Thomas Dziubla (University Of Kentucky); Kevin Baldridge (University Of Kentucky)

2:30 pm (Abbey South, Monday)

Oral 56 - Ionic Strength And Ph Responsive UF And RO Membranes (In-Person)

Yoram Cohen (UCLA)*; Yian Chen (UCLA)

3:00 pm (Abbey South, Monday)

Oral 57 - Catalyst-Free Water-Gas Shift Reaction In Ceramic-Carbonate Dual-Phase Membrane Reactors At High Pressures – Effect Of Side Reaction (In-Person)

Oscar Ovalle (Arizona State University); Jerry Lin (Arizona State University)*

3:30 pm (Abbey South, Monday)

Oral 58 - Catalytic Membranes And The Role Of Biohydrogen For Reductive Chloro-Organics Detoxification From Water (In-Person)

Dibakar Bhattacharyya (U. Kentucky)*; Rollie Mills (U. Kentucky); Hongyi Wan (University Of Kentucky); Michael J Detisch (University Of Kentucky); Todd Hastings (University Of Kentucky); Noah Meeks (Southern Services Co)

4:00 pm (Abbey South, Monday)

Oral 59 - Catalytic Membrane Reactor For Conversion Of Waste Biomass To Chemical Intermediates (In-Person)

Ranil Wickramasinghe (University Of Arkansas)*; Zhexi Zhu (University Of Arkansas); Xianghong Qian (University Of Arkansas)

4:30 pm (Abbey South, Monday)

Oral 60 - Trends, Mechanisms, And Opportunities In Using Electrically Responsive Membranes (Virtually)

Melissa J Larocque (Mcmaster University); Adi Gelb (Mcmaster University); David Latulippe (Mcmaster University); Charles-François De Lannoy (Mcmaster)*

Oral Sessions - Tuesday

Presenting authors are indicated by an asterisk*. See the most up-to-date version of the program book at **www.membranes.org/nams~2022/**

Oral 11 – POLYMERIC AND MIXED-MATERIALS – GAS SEPARATION I

9:30 am – 12:30 pm, ROOM Palm AD

Session Chairs
Benjamin J. Sundell
C2Sense, Inc

Zachary Smith
Massachusetts Institute of Technology

Xiaoli Ma
University of Wisconsin-Milwaukee

Raj Singh
Los Alamos National Lab

9:30 am (Palm AD, Tuesday)

Oral 61 - Unexpected Size-Controlled Vapor Sorption In Glassy Membranes Exhibiting Configurational Free Volume (In-Person)

William J. Box (University Of Oklahoma)*; Zihan Huang (University Of Notre Dame); Ruilan Guo (University Of Notre Dame); Michele Galizia (University Of Oklahoma)

10:00 am (Palm AD, Tuesday)

Oral 62 - Microporous Polymers With Tailored Cavities By Direct Fluorination For Small Gas Molecule Separations (In-Person)

Wonhee Lee (Georgia Institute Of Technology)*; Yi Ren (Georgia Institute Of Technology); Conrad J Roos (Georgia Institute Of Technology); Ryan Lively (Georgia Tech); Jong Geun Seong (Hanyang University); Jongmyeong Lee (Hanyang University); So Young Lee (Hanyang University); Yuseong Do (Hanyang University); Joon Yong Bae (Hanyang University); Sun Ju Moon (Hanyang University); Chi Hoon Park (Hanyang University); Hyejin Jo (Hanyang University); Ju Sung Kim (Hanyang University); Young Moo Lee (Hanyang University); Kueir-Rarn Lee (Chung Yuan University); Wei-Song Hung (Chung Yuan University); Juin-Yih Lai (Chung Yuan University)

10:30 am (Palm AD, Tuesday)

Oral 63 - PIMs at Nanoscale (In-person)

Wojciech Ogieglo (KAUST)*; Ingo Pinnau (KAUST)

11:00 am (Palm AD, Tuesday)

Oral 64 - Polyformamidine As Fixed-Site Carrier For CO_2/N_2 Separation Membranes (In-Person)

Jingying Hu (The Ohio State University)*; Yang Han (The Ohio State University); Winston Ho (The Ohio State University)

11:30 am (Palm AD, Tuesday)

Oral 65 - Scalable One-Step Fabrication Of MOF-Based Asymmetric Mixed-Matrix Membranes With Exceptionally High Propylene/Propane Separation Performance (In-Person)

Yinying Hua (Texas A&M University)*; Sunghwan Park (Johns Hopkins University); Hae-Kwon Jeong (Texas A&M University)

12:00 pm (Palm AD, Tuesday)

Oral 66 - Effect Of PPE Decontamination Processes On The Filtration Efficiency Of Porous Polysulfone Flat Sheet Membrane (In-Person)

Ebuka Ogbuoji (University Of Kentucky)*; Isabel Escobar (University Of Kentucky)

Oral 12 – BIOINSPIRED AND BIOMIMETIC MATERIALS

9:30 am – 12:30 pm, ROOM Palm BE

Session Chairs
Yuexiao Shen
Texas Tech University

Cassandra Porter
Auburn Univeristy

9:30 am (Palm BE, Tuesday)

Oral 67 - Artificial Water Channels-Towards Biomimetic Membranes For Desalination (In-Person)

Mihail Barboiu* (Institut Europeen Des Membranes).

10:00 am (Palm BE, Tuesday)

Oral 68 - Quantification Of Salt And Proton Rejection In A Flourofoldamer Based Artificial Water Channel (In-Person)

Laxmicharan Samineni (University Of Texas At Austin)*; Matthew Skiles (University Of Texas At Austin); Woochul Song (University Of Texas At Austin); Manish Kumar (University Of Texas At Austin)

10:30 am (Palm BE, Tuesday)

Oral 69 - Fast Water Transport Through Biomimetic Reverse Osmosis Membranes Embedded With Peptide-Attached (Pr)-Pillar[5]Arenes Water Channels (In-Person)

Yu Jie Lim (Nanyang Technological University)*; Kunli Goh (Nanyang Technological University); Gwo Sung Lai (Nanyang Technological University); Chiann Yi Ng (Nanyang Technological University); Jaume Torres (Nanyang Technological University); Rong Wang (Nanyang Technological University)

11:00 am (Palm BE, Tuesday)

Oral 70 - Preparation Of Capillary Flow Membranes From Wood Veneers With Advanced Electron And Ion Microscopy Characterization (Virtually)

Nadezda Prochukhan (Trinity College Dublin)*; Lucy Fitzsimmons (Trinity College Dublin); Michael Morris (Trinity College Dublin)

11:30 am (Palm BE, Tuesday)

Oral 71 - Skin Like Biomimetic Membranes For Breathable Protective Fabrics (In-Person)

Hyeonji Oh (University Of Texas At Austin)*; Yu-Ming Tu (The University Of Texas At Austin); Benny Freeman (University Of Texas At Austin); Manish Kumar (University Of Texas At Austin)

12:00 pm (Palm BE, Tuesday)

Oral 72 - A TiO_2-Au Janus Membrane Has Potential To Enable Self-Pumping Flow By Light. (In-Person)

Yuhang Fang (Purdue University)*; Jefferey Moran (George Mason University); David M Warsinger (Purdue)

Oral 13 – MEMBRANES FOR ELECTROCHEMICAL APPLICATIONS I

9:30 am – 12:30 pm, ROOM Palm CF

Session Chairs
Hee Jeung Oh
Penn State University

Piran R Kidambi
Vanderbilt University

9:30 am (Palm CF, Tuesday)

Oral 73 - Deconstructing Proton Transport Through Atomically Thin Monolayer CVD Graphene Membranes (In-Person)

Piran Kidambi (Vanderbilt University)*

10:00 am (Palm CF, Tuesday)

Oral 74 - Investigation Of Porous PVDF Membranes Behavior During Metal-Assisted Chemical Imprinting (In-Person)

Aliaksandr Sharstniou (Arizona State University); Shouhong Fan (University Of Colorado At Boulder); Emmanuel Dasinor (Arizona State University); Yifu Ding (Membrane Science, Engineering, And Technology (MAST) Center, Department Of Mechanical Engineering, University Of Colorado Boulder); Bruno Azaredo (Arizona State University)*

10:30 am (Palm CF, Tuesday)

Oral 75 - Dialysate Regeneration Via Photooxidation And FO Membrane Protection-Loop To Enable Portable Kidney Dialysis (Virtually)

Guozheng Shao (Univ. Of Wash.); Bruce Hinds (Univ. Of Washington)*

11:00 am (Palm CF, Tuesday)

Oral 76 - Electrochemical Oxygen Pumps For Mobile Medical Oxygen (In-Person)

Ralph A Bauer (Global Research And Development Inc.)*

11:30 am (Palm CF, Tuesday)

Oral 77 - Liquid Separations Membranes Made Of Functionalized Poly-Ether Ether Ketone Blended With Polysulfone (In-Person)

Abelline K Fionah (University Of Kentucky)*; Laura Brady (University Of Kentucky); Kate Mclarney (University Of Kentucky); Isabel Escobar (University Of Kentucky)

12:00 pm (Palm CF, Tuesday)

Oral 78 - Lithium Brine Purification And Electrochemical Conversion Via Selective Bipolar Membrane Electrodialysis (In-Person)

Aubrey Quigley (University Of Texas At Austin); Kevin Reimund (University Of Texas At Austin)*; Benjamin Pedretti (University Of Texas At Austin); Kristofer Gleason (University Of Texas At Austin); Nathaniel Lynd (The University Of Texas At Austin); Benny Freeman (The University Of Texas At Austin)

Oral 14 – HIGH SALINITY STREAMS, BRINE MINIMIZATION AND ZLD

9:30 am – 12:30 pm, ROOM Abbey South

Session Chairs
Kerri Hickenbottom
The University of Arizona

Jonathan A. Brant
University of Wyoming

9:30 am (Room Abbey South, Tuesday)

Oral 79 - High Pressure Reverse Osmosis Membrane Modeling And Characterization (In-Person)

Jeffrey R Mccutcheon (University Of Connecticut)*; Yara Suleiman (University Of Connecticut); Danh Nguyen (University Of Connecticut); Mi Zhang (University Of Connecticut); Ramesh Bhave (Ornl); Ramanan Sankaran (Oak Ridge National Laboratory); Luke Malencica (Oak Ridge National Laboratory); Kyle Cole (Michigan State University); Mayur Ostwal (University Of Connecticut); Sina Shahbazmohamadi (University Of Connecticut); Menachem Elimelech (Yale University); Ying Li (University Of Connecticut); Marcus Day (National Renewable Energy Laboratory)

10:00 am (Room Abbey South, Tuesday)

Oral 80 - Selecting Fouling And Scaling Mitigation Strategies For Membrane Distillation Applications With Water Reuse Reverse Osmosis Concentrate (In-Person)

Luke Presson (University Of Arizona)*; Mukta Hardikar (The University Of Arizona); Varinia V Felix (The University Of Arizona); Andrea Achilli (The University Of Arizona); Kerri Hickenbottom (U. Arizona)

10:30 am (Room Abbey South, Tuesday)

Oral 81 - Highly Charged Ion-Exchange Membranes For Treatment Of Brines Via Electrodialysis (In-Person)

Carolina Espinoza (University Of Michigan)*; José Carlos Díaz (University Of Michigan); Kevin Kim

(University Of Michigan); Jovan Kamcev (University Of Michigan)

11:00 am (Room Abbey South, Tuesday)

Oral 82 - Characterizing Low-Salt-Rejection RO Membranes With The Kedem-Katchalsky Model (Virtually)

Abdessamad Belgada (Yale University)*; Yuhao Du (Yale University); Li Wang (Yale University); Menachem Elimelech (Yale University)

11:30 am (Room Abbey South, Tuesday)

Oral 83 - Screening For Scaling Resistant Desalination Membranes (In-Person)

Sankaranarayanan Ayyakudi Ravichandran (University Of Colorado Boulder); Saied Delagah (U.S. Department Of Interior); John Pellegrino (University Of Colorado)*

12:00 pm (Room Abbey South, Tuesday)

Oral 84 - Optimizing Pore Size And Charge Of Polyamide Nanofiltration Membrane For Lithium Enrichment From Salt Lake Brine (In-Person)

Juan Zhai (Texas Tech University)*; Sunday P Adu (Texas Tech University); Yuexiao Shen (Texas Tech University)

Oral 15 – PROCESS INTENSIFICATION AND INTEGRATION

9:30 am – 12:30 pm, ROOM Abbey North

Session Chairs

Henry Jonathan Tanudjaja
Nanyang Technological University

Marie-Eve Langevin
Ameridia Innovative Solutions Inc

Mahdi Malmali
Texas Tech University

Albert S. Kim
University of Hawai'i

9:30 am (Room Abbey North, Tuesday)

Oral 85 - Assessment Of A Hybrid Forward Osmosis-Freeze Crystallization Process For Liquid Mining Of Lithium: Experimental Study And Numerical Simulation (Virtually)

Afshin Amani (Université Laval)*; Georgios Kolliopoulos (Université Laval)

10:00 am (Room Abbey North, Tuesday)

Oral 86 - Thermal Storage Management Integrated with Solar Driven Membrane Distillation for Process Intensification (In-person)

Mikah M Inkawhich (University of Arizona); Jeb C Shingler (University of Arizona); Spencer Ketchum (University of Arizona); Wei Pan (DWP Energy Solutions); Robert Norwood (University of Arizona); Andrea Achilli (The University of Arizona); Kerri Hickenbottom (University of Arizona)*

10:30 am (Room Abbey North, Tuesday)

Oral 87 - Reverse Osmosis vs. Electrodialysis: Comparing The Cost Of Brackish Water Desalination (In-Person)

Sohum Patel (Yale University)*; Boreum Lee (Yale University); Menachem Elimelech (Yale University)

11:00 am (Room Abbey North, Tuesday)

Oral 88 - Biocidal Effects In Microwave Induced Membrane Distillation (Virtually)

Indrani Gupta (New Jersey Institute Of Technology)*; Sagar Roy (New Jersey Institute Of Technology); Somenath Mitra (New Jersey Institute Of Technology)

11:30 am (Room Abbey North, Tuesday)

Oral 89 - Automated And Field-Deployable Evaluation Of Microfiltration/Ultrafiltration Fouling (In-Person)

Weiming Qi (Clemson University)*; David Ladner (Clemson)

12:00 pm (Room Abbey North, Tuesday)

Oral 90 - Scale-Up Of Membrane Distillation For Treating Produced Waters From Oil And Gas Industry. (In-Person)

Ritesh Pawar (University Of Pittsburgh)*; Zhewei Zhang (University Of Pittsburgh); Radisav Vidic (University Of Pittsburgh)

Oral Sessions - Tuesday

Presenting authors are indicated by an asterisk*. See the most up-to-date version of the program book at **www.membranes.org/nams~2022/**

Oral 16 – POLYMERIC AND MIXED-MATERIALS – GAS SEPARATION II

2:00 pm – 5:00 pm, ROOM Palm AD

Session Chairs

Benjamin J. Sundell
C2Sense, Inc

Zachary Smith
Massachusetts Institute of Technology

Xiaoli Ma
University of Wisconsin-Milwaukee

Raj Singh
Los Alamos National Lab

2:00 pm (Palm AD, Tuesday)

Oral 91 - Reduction Of Physical Aging In Copper Nanoparticle Pillared CMSMs Derived From A Polyimide Precursor Incorporating The Bis(Phenyl)Fluorene-Based Cardo Moiety And Uniformly Dispersed Metal Organic Polyhedra-18 (MOP-18) (In-Person)

Masoumeh Tajik Asl (The University Of Texas At Dallas)*; John Ferraris (University Of Texas At Dallas); Kenneth Balkus (Univerisity Of Texas At Dallas); Inga Musselman (University Of Texas At Dallas)

2:30 pm (Palm AD, Tuesday)

Oral 92 - Novel Graphene Oxide-Based Membrane Structure For A Highly Effective Breathable Barrier For Toxic Vapors And Chemical Warfare Agents (In-Person)

Yufeng Song (New Jersey Institute Of Technology); Cheng Peng (New Jersey Institute Of Technology); Zafar Iqbal (New Jersey Institute Of Technology); Kamalesh K Sirkar (New Jersey Institute Of Technology)*; Gregory Peterson (4CBR Filtration Branch, R&T Directorate DEVCOM Chemical Biological Center, U.S. Army Futures Command)

3:00 pm (Palm AD, Tuesday)

Oral 93 - Peculiar Effect Of Low Loading Of Metal-Organic Polyhedra On CO_2/N_2 Separation Properties Of Cross-Linked Polyethers (In-Person)

Taliehsadat Alebrahim (The State University Of New York At Buffalo)*; Liang Huang (The State University Of New York At Buffalo); Heshali Welgama (The State University Of New York At Buffalo); Shiwang Cheng (Michigan State University); Durga Acharya (CSIRO); Cara Doherty (CSIRO); Anita Hill (CSIRO); Clayton Rumsey (The State University Of New York At Buffalo); Martin Trebbin (The State University Of New York At Buffalo); Tim Cook (University At Buffalo); Haiqing Lin (The State University Of New York At Buffalo)

3:30 pm (Palm AD, Tuesday)

Oral 94 - Cross-Linked Poly(Ionic Liquid)-Ionic Liquid Composite Membranes For CO_2/Light Gas Separations: Mixed-Gas Separation Performance And Long-Term Stability Under High-Temperature And -Pressure Conditions (Virtually)

Chamaal Karunaweera (University Of Colorado Boulder)*; Douglas Gin (Department Of Chemistry, University Of Colorado - Boulder); Richard Noble (Deparment Of Chemical And Biological Engineering, University Of Colorado - Boulder)

4:00 pm (Palm AD, Tuesday)

Oral 95 - Directing Flexibility Of Metal-Organic Framework Toward Crystallographically Derived Molecular Sieving (In-Person)

Ke Zhang (Aramco Americas)*

4:30 pm (Palm AD, Tuesday)

Oral 96 - Ethylene And Ethane Transport Properties Of Hydrogen-Stable Ag^+-Based Facilitated Transport Membranes (In-Person)

Matthew N Davenport (The University Of Texas At Austin)*; Caitlin Bentley (The University Of Texas At Austin); Joan Brennecke (The University Of Texas At Austin); Benny Freeman (The University Of Texas At Austin)

Oral 17 – DOWNSTREAM BIOPROCESSING

2:00 pm – 5:00 pm, ROOM Palm BE

Session Chairs
Onur Kas
University of California, Berkeley

James McGrath
University of Rochester

2:00 pm (Palm BE, Tuesday)

Oral 97 - Membrane Adsorber Purification Of Pdna (Virtually)

Herb Lutz (Milliporesigma)*; Thomas Elich (Milliporesigma); Tyler Cheung (Milliporesigma)

2:30 pm (Palm BE, Tuesday)

Oral 98 - Predicting The Transport Of Soft Droplets In Porous Media From Measurable Emergent Properties (In-Person)

Guillaume Lostec (University Of Colorado-Boulder); Franck Vernerey (University Of Colorado-Boulder); John Pellegrino (University Of Colorado)*

3:00 pm (Palm BE, Tuesday)

Oral 99 - Purification Of Viral Vectors With Novel Membrane Adsorbers (In-Person)

Cristiana Boi (Università Di Bologna); Jinxin Fan (North Carolina State University)*; Ruben Carbonell (North Carolina State University)

3:30 pm (Palm BE, Tuesday)

Oral 100 - The Effects Of Low Flow And Flow Disruption On Virus Filtration (Virtually)

Wenbo Xu (University Of Arkansas); Ranil Wickramasinghe (University Of Arkansas); Xianghong Qian (University Of Arkansas)*

4:00 pm (Palm BE, Tuesday)

Oral 101 - High Purification Of Binary Protein Mixtures Having Close Molecular Weights By Ultrafiltration (In-Person)

Lixin Feng (New Jersey Institute Of Technology); Yufeng Song (New Jersey Institute Of Technology)*; Sagnik Basuray (New Jersey Institute Of Technology); Kamalesh K Sirkar (New Jersey Institute Of Technology); Solomon Isu (University Of Arkansas); Ranil Wickramasinghe (University Of Arkansas)

4:30 pm (Palm BE, Tuesday)

Oral 102 - Understanding The Mechanisms That Govern B. Diminuta Microfiltration To Improve The Sterile Filtration Of Therapeutic Viruses (In-Person)

Evan Wright (McMaster University)*; Alexandra Jucan (McMaster University); David Latulippe (McMaster University)

Oral 18 – MEMBRANES FOR ELECTROCHEMICAL APPLICATIONS II

2:00 pm – 5:00 pm, ROOM Palm CF

Session Chairs
Hee Jeung Oh
Penn State University

Piran R Kidambi
Vanderbilt University

2:00 pm (Palm CF, Tuesday)

Oral 103 - Separation Of Rare Earth Elements From Simulated Geothermal Water Using Novel Electrodialysis Metathesis Process (In-Person)

Lucy M Camacho (Texas A&M University-Kingsville)*; Mohammad U. Shafiq (Texas A&M University-Kingsville)

2:30 pm (Palm CF, Tuesday)

Oral 104 - Electrocatalytic Upcycling Of Nitrate Wastewater To Ammonia Fertilizer Via Electrified Membrane (In-Person)

Jianan Gao (New Jersey Institute Of Technology)*; Yifan Li (Qingdao University Of Technology); Bo Jiang (Qingdao University Of Technology); Taha Marhaba (New Jersey Institute Of Technology); Wen Zhang (New Jersey Institute Of Technology)

3:00 pm (Palm CF, Tuesday)

Oral 105 - Co-Transport Of Lithium, Sodium, And Potassium Ions In Polysulfone-Based Cems (In-Person)

Jung Min Kim (University Of Virginia)*; Sean M Bannon (University Of Virginia); Geoffrey M Geise (University Of Virginia)

3:30 pm (Palm CF, Tuesday)

Oral 106 - Zwitterions And Their Influence Block Polymer Morphology (In-Person)

Bradley Grim (Arizona State University)*; Matthew Green (Arizona State University); Frederick Beyer (Army Research Laboratory)

4:00 pm (Palm CF, Tuesday)

Oral 107 - Recent Progress In High Capacity, Stable, Lithium-Sulfur Batteries. (In-Person)

Matthew R Hill (CSIRO And Monash University)*

4:30 pm (Palm CF, Tuesday)

Oral 108 - Diffusion Coefficients Of Condensed Counterions In Ion-Exchange Membranes: Application Of Screening Length Scaling Relationship In Concentrated Electrolytes (In-Person)

Yuxuan Huang (Columbia University)*; Hanqing Fan (Columbia University); Ngai Yin Yip (Columbia University)

Oral 19 – MEMBRANE CHARACTERIZATION

2:00 pm – 5:00 pm, ROOM Abbey South

Session Chairs
Weiyi Li
Southern University of Science and Technology

Santiago Romero
The University of Edinburgh

Yang Liu
Aramco Americas

2:00 pm (Room Abbey South, Tuesday)

Oral 109 - Multi-Lab Study On The Pure-Gas Permeation Of Commercial Polysulfone (PSF) Membranes: Measurement Standards And Best Practices (In-Person)

Katherine Mizrahi Rodriguez (MIT)*; Wan-Ni Wu (MIT); Zachary Smith (MIT)

2:30 pm (Room Abbey South, Tuesday)

Oral 110 - A Diafiltration Apparatus For High-Throughput Analysis Of Membrane Transport Properties (In-Person)

Jonathan Ouimet (University Of Notre Dame); Xinhong Liu (University Of Notre Dame); Zachary Muetzel (University Of Notre Dame); David Brown (University Of Notre Dame); Elvis Eugene (University Of Notre Dame); Tylar Popps (University Of Notre Dame); Alexander Dowling (University Of Notre Dame); Bill Phillip (Notre Dame)*

3:00 pm (Room Abbey South, Tuesday)

Oral 111 - New Insights in Ion Transport in Polyamide Membranes Using Impedance Spectroscopy (Virtually)

Viatcheslav Freger (Technion - IIT)*; Mikhail Stolov (Technion - Israel Institute of Technology)

3:30 pm (Room Abbey South, Tuesday)

Oral 112 - Interferometry For Precisely Measuring Ultralow Flow Rates From Permeable Materials (Virtually)

Cody Ritt (Yale University)*

4:00 pm (Room Abbey South, Tuesday)

Oral 113 - The Influence Of Amine Structure And Water On The Mechanism Of CO_2 Facilitated Transport: Operando Characterization And Kinetic Modeling (In-Person)

Casey O'Brien (University Of Notre Dame)*

4:30 pm (Room Abbey South, Tuesday)

Oral 114 - Inadequacy Of Current Approaches For Characterizing Membrane Transport Properties At High Salinities (In-Person)

Yuanzhe Liang (Stanford University)*; Alexander Dudchenko (SLAC National Accelerator Laboratory); Meagan Mauter (Stanford Univerisity)

Oral 20 – MOLECULAR AND PROCESS MODELING

2:00 pm – 5:00 pm, ROOM Abbey North

Session Chairs

David M. Warsinger
Purdue University

Christina Carbrello
MilliporeSigma

Shihong Lin
Vanderbilt University

2:00 pm (Room Abbey North, Tuesday)

Oral 115 - The Impact Of Ion-Ion Correlated Motion On Salt Transport In Solvated Ion Exchange Membranes (Virtually)

Nico Marioni (The University Of Texas At Austin)*; Zidan Zhang (The University Of Texas At Austin); Harnoor Sachar (The University Of Texas At Austin); Everett S Zofchak (The University Of Texas At Austin); Sanket Kadulkar (The University Of Texas At Austin); Tyler J Duncan (The University Of Texas At Austin); Benny Freeman (The University Of Texas At Austin); Venkat Ganesan (The University Of Texas At Austin)

2:30 pm (Room Abbey North, Tuesday)

Oral 116 - Elucidating CO_2/Hydrocarbons Solubility Selectivity In Aliphatic Polycarbonates Via DFT Calculation And Experimentation (In-Person)

Thien N Tran (University At Buffalo, The State University Of New York)*; Yuqing Fu (University Of California, Riverside); De-En Jiang (University Of California, Riverside); Haiqing Lin (The State University Of New York At Buffalo)

3:00 pm (Room Abbey North, Tuesday)

Oral 117 - Non-Equilibrium Statistical Mechanics Approach For Extended Solution-Diffusion Model (Virtually)

Albert Kim (U. Of Hawai'i)*

3:30 pm (Room Abbey North, Tuesday)

Oral 118 - Mechanisms Of Ion And Water Transport In PEGDA Membranes (Virtually)

Everett S Zofchak (The University Of Texas At Austin)*; Zidan Zhang (The University Of Texas At Austin); Harnoor Sachar (The University Of Texas At Austin); Nico Marioni (The University Of Texas At Austin); Sanket Kadulkar (The University Of Texas At Austin); Tyler J Duncan (The University Of Texas At Austin); Benny Freeman (The University Of Texas At Austin); Venkat Ganesan (The University Of Texas At Austin)

4:00 pm (Room Abbey North, Tuesday)

Oral 119 - Decoding The Molecular-Scale Determinants Of Antifouling At Polymer Membrane Selective Layers (In-Person)

Dennis C Robinson Brown (The University Of California Santa Barbara)*; Joshua Moon (California Nanosystems Institute, The University Of California Santa Barbara); Thomas Webber (The University Of California Santa Barbara); Lynn Katz (The University Of Texas At Austin); Rachel Segalman (The University Of California Santa Barbara); Scott Shell (The University Of California Santa Barbara); Songi Han (The University Of California Santa Barbara)

4:30 pm (Room Abbey North, Tuesday)

Oral 120 - Towards A Universal Framework For Evaluating Mass Transport In Pressure, Concentration, And Temperature Driven Membrane-Based Desalination Systems (In-person)

Kian P Lopez (University Of Colorado Boulder)*; Ruoyu Wang (Vanderbilt University); Elizabeth A Hjelvik (University Of Colorado Boulder); Shihong Lin (Vanderbilt University); Anthony Straub (University Of Colorado Boulder)

Oral Sessions - Wednesday

Presenting authors are indicated by an asterisk*. See the most up-to-date version of the program book at www.membranes.org/nams~2022/

Oral 21 – AWARDS SESSION

9:30 am – 12:30 pm, ROOM Palm AD

Session Chairs
Caleb Funk
DuPont

9:30 am (Palm AD, Wednesday)

Oral 121 - Diglycolamide Membrane Adsorbers To Separate Lanthanides And Actinides For Use In Radiopharmaceuticals (In-Person)

Christine E Duval (Case Western Reserve University)*; Maura Sepesy (Case Western Reserve University); Alec Johnson (Case Western Reserve University); Tuli Banik (Case Western Reserve University); Megan Sibley (Case Western Reserve University); Joelle V Scott (Case Western Reserve University); Alexa Ford (Case Western Reserve University); Timothy Yen (Case Western Reserve University)

10:00 am (Palm AD, Wednesday)

Oral 122 - Putting Bubbles To Work: Next-Generation Water Treatment Systems Using Air-Trapping Membranes (In-Person)

Anthony Straub (University Of Colorado Boulder)*

10:30 am (Palm AD, Wednesday)

Oral 123 - Elucidating The Role Of Micropore Generating Backbone Motifs And Amine Functionality On Sorption Energetics, Mixed-Gas Permeation, And H_2S Transport (In-Person)

Katherine Mizrahi Rodriguez (MIT)*; Pablo Dean (MIT); Naksha Roy (MIT); Sheng Guo (MIT); Timothy Swager (MIT); Zachary Smith (MIT)

11:00 am (Palm AD, Wednesday)

Oral 124 - Selective Ion-Ion Separations And Desalination Using Adsorptive Ion-Capture Electrodialysis Membranes (Virtually)

Adam A Uliana (University Of California, Berkeley)*; Ngoc Bui (The University Of Oklahoma); Jovan Kamcev (University Of Michigan); Mercedes Taylor (University Of Maryland); Jeffrey Urban (Lawrence Berkeley National Laboratory); Jeffrey Long (University Of California, Berkeley)

11:30 am (Palm AD, Wednesday)

Oral 125 - Engineering Subnanometer Vacancy Defects In Single-Layer Graphene For High-Performance Gas Separations (In-Person)

Luis Francisco Villalobos (Yale University)*; Kumar Varoon Agrawal (Ecole Polytechnique Fédérale De Lausanne)

12:00 pm (Palm AD, Wednesday)

Oral 126 - Designing Cation-Exchange Membranes With Coordination-Based Selectivity Between Cations (In-Person)

Ryan Duchanois (Yale University)*; Lauren A Mazurowski (Yale University); Menachem Elimelech (Yale University)

Oral 22 - MEMBRANE FOULING I

9:30 am – 12:30 pm, ROOM Palm CF

Session Chairs
Steven Weinman
The University of Alabama

Ngoc T. Bui
The University of Oklahoma

Dan Miller
Lawrence Berkeley National Laboratory

Audie Thompson
University of Arkansas

9:30 am (Palm CF, Wednesday)

Oral 127 - A Computational And Experimental Test Bed For Prediction Of RO Module Fouling (In-Person)

Daniel J Miller (Lawrence Berkeley National Laboratory)*; Sergi Molins Rafa (Lawrence Berkeley National Laboratory); Mostafa Nassr (UT Austin); Sarah Dischinger (Lawrence Berkeley National Laboratory); Nicolas Spycher (LBNL); Jacob Johnston (Colorado School Of Mines); Ji Yeon Lee (Lawrence Berkeley National Laboratory); Pedram Bigdelou (Colorado School Of Mines); Kristofer Gleason (The University Of Texas At Austin); Nils Tilton (Colorado School Of Mines); William Stringfellow (Lawrence Berkeley National Laboratory); Benny Freeman (The University Of Texas At Austin)

10:00 am (Palm CF, Wednesday)

Oral 128 - Application Of Machine Learning-Based Models To Understand And Predict Critical Flux Of Oil-In-Water Emulsion In Crossflow Microfiltration (In-Person)

Henry J Tanudjaja (Nanyang Technological University)*; Jia Wei Chew (Nanyang)

10:30 am (Palm CF, Wednesday)

Oral 129 - Quantifying Electrically Responsive Membranes For Biofouling Mitigation (Virtually)

Charles-François De Lannoy (Mcmaster)*; Amin Halali (Mcmaster University)

11:00 am (Palm CF, Wednesday)

Oral 130 - Feed Temperature Effects on Organic Fouling of Reverse Osmosis Membranes: Competition of Interfacial and Transport Properties (In-person)

Santiago Romero-Vargas Castrillon (University of Edinburgh)*

11:30 am (Palm CF, Wednesday)

Oral 131 - Acoustically Excited Microstructure For On-Demand Fouling Mitigation In A Microfluidic Membrane Filtration Device (In-Person)

Kieran R Fung (University Of Colorado Boulder)*; Yuekang Li (University Of Colorado Boulder); Shouhong Fan (University Of Colorado Boulder); Apresio Fajrial (University Of Colorado Boulder); Yifu Ding (University Of Colorado Boulder); Xiaoyun Ding (University Of Colorado Boulder)

12:00 pm (Palm CF, Wednesday)

Oral 132 - Unraveling The Mechanism Of An AC Electric Field Applied For Fouling Mitigation (Virtually)

Saikat Bhattacharjee (Technion-Israel Institute Of Technology)*; Guy Z Ramon (Technion - Israel Institute Of Technology)

Oral 23 – MEMBRANE DISTILLATION AND PERVAPORATION

9:30 am – 12:30 pm, ROOM Abbey North

Session Chairs

Lucy Mar. Camacho
Texas A&M University, Kingsville

Kailong Jin
Arizona State University

9:30 am (Abbey North, Wednesday)

Oral 133 - Vapor Flux Of Membrane Distillation: Theoretical Limits, Insights On Membrane Design, And Anomalous Phenomena (In-Person)

Ruoyu Wang (Vanderbilt University)*; Shihong Lin (Vanderbilt University)

10:00 am (Abbey North, Wednesday)

Oral 134 - Modeling Pilot-Scale Membrane Distillation At High Salinity Using Bench-Scale Data (In-Person)

Mukta Hardikar (The University Of Arizona)*; Itzel Marquez (Central Michigan University); Eduardo

Saez (University Of Arizona); Andrea Achilli (The University Of Arizona)

10:30 am (Abbey North, Wednesday)

Oral 135 - Integrated Electrocoagulation, Membrane Filtration And Membrane Distillation For Treatment Of Hydraulic Fracturing Produced Water (In-Person)

Ranil Wickramasinghe (University Of Arkansas)*; Mahmood Jebur (University Of Arkansas); Yuhsuan Chiao (University Of Arkansas)

11:00 am (Abbey North, Wednesday)

Oral 136 - Fermentation Broth Recovery Of 2,3 Butanediol Using Polymer Based Ceramic Membrane For Biofuel Production (Virtually)

Oindrila Gupta (Oak Ridge National Laboratory)*; Aimee Church (Oak Ridge National Laboratory)

11:30 am (Abbey North, Wednesday)

Oral 137 - Impact Of Oxidative Chemicals On The Performance And Materials Properties Of Hydrophobic Porous Membranes Used In Membrane Distillation (In-Person)

Hannah Cairney (University Of Colorado Boulder); Elizabeth A Hjelvik (University Of Colorado Boulder)*; Duong T. Nguyen (University Of Colorado Boulder); Anthony Straub (University Of Colorado Boulder)

12:00 pm (Abbey North, Wednesday)

Oral 138 - Porous Metal Foam Condensation Doubles Energy Efficiency Of Membrane Distillation (In-Person)

Hamid Fattahijuybari (Purdue University)*; Harsharaj B. Parmar (Purdue University); Ali D. Alshubbar (Purdue University); Katherine L. Young (Purdue University); David M Warsinger (Purdue)

Oral 24 – INDUSTRIAL APPLICATIONS

9:30 am – 12:30 pm, ROOM Palm BE

Session Chairs

Dibakar Bhattacharyya
University of Kentucky

CJ Kurth
Aqua Membranes, Inc

Evan Hatakeyama
Chevron

9:30 am (Palm BE, Wednesday)

Oral 139 - An Enterprise Approach To Developing Industrial Membrane-Based Solutions (In-Person)

Adil M Dhalla (Separation Technologies Applied Research And Translation (START) Centre, Nanyang Technological University - Ntuitive Pte Ltd)*; Chakravarthy Gudipati (START - Ntuitive Pte Ltd); Joanne Sim (SG MEM, Singapore Membrane Consortium, Nanyang Technological University - Ntuitive Pte Ltd)

10:00 am (Palm BE, Wednesday)

Oral 140 - Single Pass TFF For Monoclonal Antibody Processing – Advancing A New Technology Through The MAST Center (In-Person)

Andrew Zydney (Penn State University)*

10:30 am (Palm BE, Wednesday)

Oral 141 - Membranes Beyond Water - Deploying Spirals In Industrial Separations (In-Person)

Benjamin Weaver (Solecta)*

11:00 am (Palm BE, Wednesday)

Oral 142 - Separation And Recovery Of Critical Materials From End-Of-Life Lithium Ion Batteries Using Membrane Solvent Extraction Process (In-Person)

Syed Islam (Oak Ridge National Laboratory)*; Priyesh Wagh (Oak Ridge National Laboratory); Ramesh Bhave (Oak Ridge National Laboratory); John Klaehn (Idaho National Laboratory)

11:30 am (Palm BE, Wednesday)

Oral 143 - Industrially Relevant Academic Research – Answering Fundamental Questions At The Module Scale (In-Person)

Jeffrey R Mccutcheon (University Of Connecticut)*; Noah Ferguson (University Of Connecticut); Maqsud Chowdhury (University Of Connecticut); Johann Schnittger (Fraunhofer-Institut Für Keramische Technologien Und Systeme IKTS); Marianne Pemberton (University Of Connecticut); Edward Wazer (University Of Connecticut); Mayur Ostwal (University Of Connecticut); Marcus Weyd (Fraunhofer-Institut Für Keramische Technologien Und Systeme IKTS); Joachim Koschikowski (Fraunhofer-Institute For Solar Energy Systems ISE)

12:00 pm (Palm BE, Wednesday)

Oral 144 - Facilitated Transport Membranes For Low Energy And Low Cost Carbon Capture And Sequestration (In-Person)

Christine Parrish (Compact Membrane Systems)*; Sudip Majumdar (Compact Membrane Systems); Ken Loprete (Compact Membrane Systems)

Oral 25 – MATERIALS FOR ELECTROCHEMICAL APPLICATIONS

9:30 am – 12:30 pm, ROOM Abbey South

Session Chairs

William Tarpeh
Stanford University

Orlando Coronell
The University Of North Carolina At Chapel Hill

9:30 am (Abbey South, Wednesday)

Oral 145 - Pressure Induced Diffusion In Hydrated Polymers (In-Person)

Rahul Sujanani (The University Of Texas At Austin)*; Kevin Reimund (University Of Texas At Austin); Kris Gleason (The University Of Texas At Austin); Benny Freeman (The University Of Texas At Austin)

10:00 am (Abbey South, Wednesday)

Oral 146 - Kinetic Control Of Intrinsic Pores In Monolayer Graphene For Large-Area Proton Selective Membranes (Virtually)

Piran Kidambi (Vanderbilt University)*

10:30 am (Abbey South, Wednesday)

Oral 147 - Exploiting The Synergistic Effect Of Surfactants And Ionic Liquids For CO_2 Capture (Virtually)

Alireza Bandegi (Case Western Reserve University); Maria Marquez Garcia (New Mexico State University); Jose Leobardo Bañuelos (The University Of Texas At El Paso); Millicent Firestone (Los Alamos National Laboratory); Reza Foudazi (University Of Oklahoma)*

11:00 am (Abbey South, Wednesday)

Oral 148 - Computational Modeling Of 3D Electrospun Nanofiber Structures (Virtually)

Albert Kim (U. Of Hawai'i)*

11:30 am (Abbey South, Wednesday)

Oral 149 - Selective Separation Of Lithium Using Chemically Modified Nanoporous Polyamide Membrane With Electrodialysis (Virtually)

Xi Chen (Stanford University)*; William Tarpeh (Stanford University)

12:00 pm (Abbey South, Wednesday)

Oral 150 - Counter-Ion Activation Energy Of Diffusion In Ion Exchange Membranes (In-Person)

David Kitto (University Of Michigan - Kamcev Lab)*; José Carlos Díaz (University Of Michigan); Carolina Espinoza (University Of Michigan); Jovan Kamcev (University Of Michigan)

Oral Sessions - Wednesday

Presenting authors are indicated by an asterisk*. See the most up-to-date version of the program book at www.membranes.org/nams~2022/

Oral 26 – SEAWATER DESAL AND OSMOTIC PROCESSES

2:00 pm – 5:00 pm, ROOM Palm BE

Session Chairs

Andrea Achilli
The University of Arizona

Milad Esfahani
The University of Alabama

Jonathan Maisonneuve
Oakland University

William Phillip
University of Notre Dame

2:00 am (Palm BE, Wednesday)

Oral 151 - Interfacial Polymerization Kinetics: New Insights On Film Formation Using In-Situ Microscopy And Particle-Tracking (In-Person)

Adi Ben Zvi (Technion)*; Guy Z Ramon (Technion - Israel Institute Of Technology)

2:30 pm (Palm BE, Wednesday)

Oral 152 - A Solution-diffusion with Defects Model to Explain Pressure-dependent Membrane Permeability in Pressure Retarded Osmosis (In-person)

Zachary Binger (The University of Arizona)*; Andrea Achilli (University of Arizona)

3:00 pm (Palm BE, Wednesday)

Oral 153 - Elucidating the Roles of Polyamide Layer Structural Properties in the Permeability-Selectivity Tradeoff Governing Aqueous Separations (Virtually)

Xi Chen (Stanford University)*; Rhea Verbeke (KU Leuven); Chanhee Boo (Korea Institute of Science and Technology); Marcel Dickmann (Universität der Bundeswehr München); Werner Egger (Universität der Bundeswehr München); Ivo Vankelecom (Universität der Bundeswehr München); Ngai Yin Yip (Columbia University)

3:30 pm (Palm BE, Wednesday)

Oral 154 - The Significance Of Membrane Charge And Co-Ion Partitioning In Determining Salt Transport Through Reverse Osmosis Membranes (Virtually)

Li Wang (Yale University)*; Kevin Pataroque (Yale University); Tianchi Cao (Yale University); Menachem Elimelech (Yale University)

4:00 pm (Palm BE, Wednesday)

Oral 155 - Forward Osmosis And Freeze Concentration For Mining And Metals Effluents (In-Person)

Noel Devaere (University Of Toronto)*; Runlin Yuan (University Of Toronto); Vladimiros Papangelakis (University Of Toronto)

4:30 pm (Palm BE, Wednesday)

Oral 156 - Predicting Efficiency And Elucidating Mechanisms For Organic And Inorganic Compound Removal By Reverse Osmosis And Nanofiltration Membranes Using Machine Learning (In-Person)

Nohyeong Jeong (Colorado State University)*; Tiezheng Tong (Colorado State University)

Oral 27 - MEMBRANE FOULING II

2:00 pm – 5:00 pm, ROOM Palm CF

Session Chairs

Steven Weinman
The University of Alabama

Ngoc T. Bui
The University of Oklahoma

Dan Miller
Lawrence Berkeley National Laboratory

Audie Thompson
University of Arkansas

2:00 pm (Palm CF, Wednesday)

Oral 157 - Fouling-Resistant Membranes With Tunable Pore Size Fabricated Using Cross-Linkable Copolymers With High Zwitterion Content (In-Person)

Samuel J Lounder (Tufts)*; Ayse Asatekin (Tufts); Luca Mazzaferro (Tufts University); Patrick Wright (Tufts University)

2:30 pm (Palm CF, Wednesday)

Oral 158 - Combined In-Situ Imaging And CFD To Unravel The Scaling Mechanism In Membrane Distillation Of High Salinity Brine (In-Person)

I-Min Hsieh (Texas Tech University); Bosong Lin (Texas Tech University); Hadis Matinpour (Texas Tech University); Mahdi Malmali (Texas Tech University)*

3:00 pm (Palm CF, Wednesday)

Oral 159 - Fouling As An Opportunity: Exploiting The Properties Of Ultrathin Membranes To Make Fouling-Based Sensors (In-Person)

James Mcgrath (University Of Rochester)*

3:30 pm (Palm CF, Wednesday)

Oral 160 - Reactive Reverse Osmosis Membranes For Dissolved Silica Antifouling And Concentration Polarization Reduction By Catalytical Degradation Of Hydrogen Peroxide (In-Person)

Weiliang Bai (University Of Texas At Austin)*; Ruizhe Xu (University Of Texas At Austin); Adam Boehnke (University Of Texas At Austin); Navid Saleh (University Of Texas At Austin); Manish Kumar (University Of Texas At Austin)

4:00 pm (Palm CF, Wednesday)

Oral 161 - Scalable Antifouling Membranes Via Zwitterionic Polymer Brushes: Examining The Roles Of Brush Thickness And Density (In-Person)

Allyson L Mcgaughey (Princeton University)*; Sneha Srinivasan (Princeton University); Kofi SS Christie (Princeton University); Zhiyong Jason Ren (Princeton University); Rodney Priestley (Princeton University)

4:30 pm (Palm CF, Wednesday)

Oral 162 - Raman Chemical Fingerprints For Real-Time Detection Of Membrane Fouling (Virtually)

Danielle Park (University Of Colorado Boulder)*; Omkar D. Supekar (University Of Colorado Boulder); Alan R. Greenberg (University Of Colorado Boulder); Juliet T. Gopinath (University Of Colorado Boulder); Victor M. Bright (University Of Colorado Boulder)

Oral 28 – CONTAMINANT REMOVAL FROM WATER

2:00 pm – 5:00 pm, ROOM Palm AD

Session Chairs

Anthony Straub
University of Colorado Boulder

Andre da Costa
Michigan Technological University

Prakhar Prakash
Chevron

Boya Xiong
University of Minnesota

2:00 pm (Palm AD, Wednesday)

Oral 163 - Development Of Highly Selective Biocatalytic Membrane Reactors For The Degradation Of Pesticides (Virtually)

Lidietta Giorno (National Research Council Of Italy - Institute On Membrane Technology (CNR-ITM))*; Giuseppe Vitola (National Research Council Of Italy, Insitute On Membrane Technology, CNR-ITM); Rosalinda Mazzei (National Research Council Of Italy, Insitute On Membrane Technology, CNR-ITM)

2:30 pm (Palm AD, Wednesday)

Oral 164 - Negative Rejection in Polyelectrolyte Multilayer Nanofiltration Membranes for Selective Ion Recovery (In-person)

Brielle Januszewski (Yale University)*

3:00 pm (Palm AD, Wednesday)

Oral 165 - Mxene-Functionalized Thin-Film Composite Membranes For PFAS Removal From Water (In-Person)

Milad Rabbani Esfahani (The University Of Alabama); Tin Le (The University Of Alabama)*

3:30 pm (Palm AD, Wednesday)

Oral 166 - Graft Polymerization Using Oppositely Charged Monomers On NF Membrane For The Efficient Removal Of Carbamazepine From Wastewater Effluents (Virtually)

Brhanu Kelali Desta (Ben-Gurion University Of The Negev)*; Akshay Modi (Ben Gurion University Of The Negev); Ron Kasher (Ben Gurion University Of The Negev)

4:00 pm (Palm AD, Wednesday)

Oral 167 - Modeling Of RO System Water Treatment Operation For Nitrate And Salt Removal Using Long-Short Term Memory (LSTM) Machine Learning Model With Attention Coefficient (In-Person)

Yoram Cohen (Professor)*; Muhammad Bilal (University of California, Los Angeles); Yang Zhou (Shanghai University)

4:30 pm (Palm AD, Wednesday)

Oral 168 - Nanofiltration And Temperature Swing Reverses Osmosis For The Removal Of Herbicide 2-Methyl-4-Chlorophenoxyacetic Acid (MCPA) From Saline Industrial Wastewater (In-Person)

Mikel Duke (Victoria University, Melbourne)*; Greg Weston (Tri-Tech Chemical Co. Pty Ltd, Australia); Stephen Gray (Victoria University); Xing Yang (Victoria University); Jianhua Zhang (Victoria University)

Oral 29 – WATER REUSE

2:00 pm – 5:00 pm, ROOM Abbey North

Session Chairs

Vicky Karanikola
The University of Arizona

Jack Girlon
Ben-Gurion University of the Negev

2:00 pm (Abbey North, Wednesday)

Oral 169 - Electroconductive Filters And Membranes: Future For Desalination And Wastewater Treatment (In-Person)

Swatantra Pratap Singh (Indian Institute Of Technology Bombay)*; Najmul Barbhuiya (Indian Institute Of Technology Bombay); Utkarsh Misra (Indian Institute Of Technology Bombay)

2:30 pm (Abbey North, Wednesday)

Oral 170 - Chlorination And Autopsy Of Fouled Engineering-Scale Reverse Osmosis Membranes (In-Person)

Bianca M. Souza Chaves (The University Of Arizona)*; Mohammed Alhussaini (The University Of Arizona); Varinia V Felix (The University of Arizona); Andrea Achilli (The University of Arizona)

3:00 pm (Abbey North, Wednesday)

Oral 171 - A Balanced-Charged Polyelectrolyte Complex Nanofiltration Membrane: Mitigating Scaling In Effluents Desalination (In-Person)

Hao Huang (Ben Gurion University Of The Negev)*; Roy Bernstein (Ben Gurion University)

3:30 pm (Abbey North, Wednesday)

Oral 172 - Degradation Of Polyamide Thin Film By Chlorine And Peracetic Acid: A Combined QCM-D And AFM Study (In-Person)

Tashfia M Mohona (University at Buffalo)*; Ning Dai (University at Buffalo); Prathima Nalam (University at Buffalo)

4:00 pm (Abbey North, Wednesday)

Oral 173 - Solute-Solute Selectivity In Membrane Separation: Definition, Measurement, And Process Scale Evaluation (In-Person)

Ruoyu Wang (Vanderbilt University)*; Shihong Lin (Vanderbilt University)

4:30 pm (Abbey North, Wednesday)

Oral 174 - Effect Of Pore Defects On Membrane Rejection In Pilot-Scale Membrane Distillation (In-Person)

Mukta Hardikar (The University Of Arizona)*; Luisa Ikner (The University Of Arizona); Varinia V Felix (The University Of Arizona); Luke Presson (University Of Arizona); Andrew Rabe (University Of Arizona); Kerri Hickenbottom (The University Of Arizona); Andrea Achilli (The University Of Arizona)

Oral 30 – MODULE MODELING AND DESIGN

2:00 pm – 5:00 pm, ROOM Abbey South

Session Chairs

David Ladner
Clemson University

Grigorios Panagakos
Carnegie Mellon University

2:00 pm (Abbey South, Wednesday)

Oral 175 - Computational Fluid Dynamics (CFD) Modeling Of 3D-Printed Spacers And Patterned Membranes In Reverse Osmosis And Nanofiltration Modules (In-Person)

David Ladner (Clemson)*; Zuo Zhou (Clemson University)

2:30 pm (Abbey South, Wednesday)

Oral 176 - Effect Of Packing Nonuniformity On The Performance Of Hollow Fiber Membrane Gas Separation Modules Fabricated From Fiber Tows (In-Person)

Lili Sun (University Of Toledo); Grigorios Panagakos (Carnegie Mellon University); Glenn Lipscomb (University Of Toledo, Toledo)*

3:00 pm (Abbey South, Wednesday)

Oral 177 - Examining The Impact Of Acid Strength On Process Efficiency And Product Usability In Ammonia Recovery From Wastewater Using Membrane Contactors (In-Person)

Abhishek Dutta (University Of British Columbia)* ; Jongho Lee (University Of British Columbia)

3:30 pm (Abbey South, Wednesday)

Oral 178 - Accelerating The Optimization Of Turbulence-Promoting Spacers Using Computational Fluid Dynamics, Surrogate Modeling, And Machine Learning (In-Person)

Zachary Binger (The University Of Arizona)*; Andrea Achilli (University Of Arizona)

4:00 pm (Abbey South, Wednesday)

Oral 179 - Flow Dynamics In A Spacer-Filled Membrane Distillation Channel (In-Person)

Bosong Lin (Texas Tech University); Hadis Matinpour (Texas Tech University); Mahdi Malmali (Texas Tech University)*

4:30 pm (Abbey South, Wednesday)

Oral 180 - Organic Matter Removal For Reverse Osmosis Fouling Mitigation: From Lab Prototype And Numerical Simulation To Pilot Plant (Virtually)

Yunqiao Ma (Nanyang Technological University)*; Tao Hua (Nanyang Technological University); Thien An Trinh (Nanyang Technological University); Rong

Wang (Nanyang Technological University); Jia Wei Chew (Nanyang)

Poster Sessions - Monday

Presenting authors are indicated by an asterisk*. See the most up-to-date version of the program book at **www.membranes.org/nams~2022/**

Poster Session (Ballroom)

7:00 pm – 10:00 pm

Session Chairs

Yuexiao Shen
Texas Tech University

Yu-Ming Tu
Massachusetts Institute Of Technology

Milad Esfahani
University Of Alabama

Poster 1 - Microcystin-LR Destruction And Removal From Water Via Enzymatic Biodegradation And Ultrafiltration

Abelline K Fionah (University Of Kentucky)*; Cannon Hacket (University Of Arkansas); Laura Brady (University Of Kentucky); Audie Thompson (Unviersity Of Arkansas); Isabel Escobar (University Of Kentucky).

Poster 2 - Chlorination And Autopsy Of End-Of-Life Engineering-Scale Reverse Osmosis Membranes

Bianca M. Souza Chaves (The University Of Arizona)*; Mohammed Alhussaini (The University Of Arizona); Varinia V Felix (The University Of Arizona); Andrea Achilli (The University Of Arizona).

Poster 3 - Quantification Of Salt And Proton Rejection In A Flourofoldamer Based Artificial Water Channel

Laxmicharan Samineni (University Of Texas At Austin)*; Matthew Skiles (University Of Texas At Austin); Woochul Song (University Of Texas At Austin); Manish Kumar (University Of Texas At Austin).

Poster 4 - Magneto-Responsive Single-Crystal $Co_0 \cdot 5Ni_0 \cdot 5FeCrO_4$: Synthesis, Dispersion Control, And Its Tunability Into Polymer Matrix Under The Augmented-Magnetic Field CO_2 Capturing And H_2 Separation Applications

Saeed Jamali Ashtiani (University Of Chemistry And Technology Prague); Zdenek Sofer (University Of Chemistry And Technology); Karel Friess (University Of Chemistry And Technology Prague)*.

Poster 5 - Optimization Of Membrane System In SMR Carbon Capture Process

David Hasse (Air Liquide)*; Noemi Collado (Air Liquide); Marcos Da Conceicao, (Air Liquide).

Poster 6 - Effect Of PPE Decontamination Processes On The Filtration Efficiency Of Flat Sheet Polysulfone Membrane

Ebuka Ogbuoji (University Of Kentucky)*.

Poster 7 - Carbon Capture From Residual Emissions Enabled By Facilitated Transport Membranes

Yang Han (The Ohio State University)*.

Poster 8 - Integrated Module Array With Facilitated Transport Membranes For Enhanced H_2 Recovery From Syngas

Yang Han (The Ohio State University)*; Yutong Yang (The Ohio State University); Ruizhi Pang (The Ohio State University); Winston Ho (OSU).

Poster 9 - Analysis On The Thermomechanical Bonding Between Membranes And Polymers

Jaylene Martinez (CU Boulder Mechanical Engineering)*; Shouhon Fan (CU Boulder Mechanical Engineering); Salil Rabade (CU Boulder Mechanical Engineering); Jason P. Killgore (Applied Chemicals And Materials Division, National Institute Of Standards And Technology, Boulder, CO); Stefano Berti (Milliporesigma); Christina Carbrello (Milliporesigma); Sean Foley (Milliporesigma); Rong Long (Membrane Science, Engineering, And Technology (MAST) Center, Department Of Mechanical Engineering, University Of Colorado Boulder); Robert Castro (Milliporesigma); Yifu Ding (Membrane Science, Engineering, And Technology

(MAST) Center, Department Of Mechanical Engineering, University Of Colorado Boulder).

Poster 10 - Air Cooled Heat Exchangers For Inland Brine Concentrate Management

Jeb C Shingler (University Of Arizona)*; Mikah M Inkawhich (University Of Arizona); Spencer Ketchum (University Of Arizona); Robert Norwood (University Of Arizona); Wei Pan (DWP Energy Solutions); Andrea Achilli (The University Of Arizona); Kerri Hickenbottom (U. Arizona).

Poster 11 - A Dialysate Free Portable Artificial Kidney Device

Leticia Santos de Souza (University of Arkansas)*; Jamie Hestekin (University Of Arkansas); Christa Hestekin (University Of Arkansas); Ira Kurtz (UCLA).

Poster 12 - Powering Up Biomedical Devices By Harnessing Energy From Reverse Electrodialysis

Jamie Hestekin (University Of Arkansas)*; Christa Hestekin (University Of Arkansas).

Poster 13 - Real-Time Vibrational Based Antifouling Mechanism For Ultrafiltration (UF) Membrane

Jasneet K Pala (The University Of Alabama)*; Milad Rabbani Esfahani (The University Of Alabama); S. Nima Mahmoodi (The University Of Alabama); Ryan Tracy (The University Of Alabama).

Poster 14 - Acoustically-Excited Microstructure For On Demand Fouling Mitigation In A Microfluidic Membrane Filtration Device

Kieran R Fung (Membrane Science, Engineering, And Technology (MAST) Center, Department Of Mechanical Engineering, University Of Colorado Boulder)*; Yuekang Li (Membrane Science, Engineering, And Technology (MAST) Center, Department Of Mechanical Engineering, University Of Colorado Boulder); Shouhong Fan (Membrane Science, Engineering, And Technology (MAST) Center, Department Of Mechanical Engineering, University Of Colorado Boulder); Apresio Fajrial (Membrane Science, Engineering, And Technology (MAST) Center, Department Of Mechanical Engineering, University Of Colorado Boulder); Yifu Ding (Membrane Science, Engineering, And Technology (MAST) Center, Department Of Mechanical Engineering, University Of Colorado Boulder); Xiaoyun Ding (Biomedical Engineering Program, Department Of Mechanical Engineering, University Of Colorado Boulder).

Poster 15 - Effect Of Nanopatterning On Concentration Polarization During Nanofiltration

Lauren M Ward (The University Of Alabama)*; Steven T Weinman (The University Of Alabama); Georgie Fickling (The University Of Alabama); Urmi Roy (The University Of Alabama).

Poster 16 - Selecting Fouling And Scaling Mitigation Strategies For Membrane Distillation Applications With Water Reuse Reverse Osmosis Concentrate (Poster)

Luke Presson (University Of Arizona)*; Mukta Hardikar (The University Of Arizona); Varinia V Felix (The University Of Arizona); Andrea Achilli (The University Of Arizona); Kerri Hickenbottom (U. Arizona).

Poster 17 - Environmental Impacts Of Desalination And Suggested Solutions

Mohamed EA Ali (Desert Research Center)*

Poster 18 - Hydrophilic Fouling-Resistant GO-ZnO/PES Membranes For Wastewater Reclamation

Oranso T Mahlangu (University Of South Africa)*; Bhekie Mamba (University Of South Africa).

Poster 19 - Air Sparged Integrated Membrane Distillation For The Treatment Of FGD Water Using Carbon Nanotube Immobilized Membrane

Mitun Chandra Bhoumick (New Jersey Institute Of Technology)*; Sagar Roy (New Jersey Institute Of Technology); Somenath Mitra (New Jersey Institute Of Technology).

Poster 20 - Membrane Mediated Degradation Of Chlorinated Organic Water Pollutants Via Metal Catalyzed Sulfate Radical Oxidation

Matthew Whitwer (University Of Kentucky)*; Rollie Mills (U. Kentucky); Lindell Ormsbee (University Of Kentucky); Dibakar Bhattacharyya (U. Kentucky).

Poster 21 - Functionalized PVDF Membranes To Control Blood Protein Behavior

Charaf Eddine Merzougui (Laboratoire De Génie Chimique, Université De Toulouse, CNRS, INPT, UPS, Toulouse, France)*; Pierre Roblin (Laboratoire De Génie Chimique, Université De Toulouse, CNRS, INPT, UPS, Toulouse, France); Patrice Bacchin (Laboratoire De Génie Chimique, Université De Toulouse, CNRS, INPT, UPS, Toulouse, France); Pierre Aimar (Laboratoire De Génie Chimique, Université De Toulouse, CNRS, INPT, UPS, Toulouse, France); Christel Causserand (Laboratoire De Génie Chimique, Université De Toulouse, CNRS, INPT, UPS, Toulouse, France).

Poster 22 - Nanofiltration As An Alternative To Reverse Osmosis As Barrier For Trace Organic Compounds In Water Reuse Applications

Mohammed Alhussaini (The University Of Arizona)*; Bianca M. Souza Chaves (The University Of Arizona); Varinia V Felix (The University Of Arizona); Andrea Achilli (The University Of Arizona).

Poster 23 - Synthesis Of Reactive Membranes For Capture And Degradation Of Water And Air Pollutants

Rollie G Mills (University Of Kentucky)*; Dibakar Bhattacharyya (U. Kentucky); Matthew D Bernard (University Of Kentucky); Ronald J. Vogler (University Of Kentucky); Jacob Concolino (University Of Kentucky); Todd Hastings (University Of Kentucky); Yinai Wei (University Of Kentucky); Thomas Dziubla (University Of Kentucky); Kevin Baldridge (University Of Kentucky).

Poster 24 - Moving Towards Cu-67 Purification With Membranes: Extracting Cu From Ni, Zn In Acidic Solution

Maura Sepesy (Case Western Reserve University)*; Benjamin Fugate (Case Western Reserve University); Christine E Duval (Case Western Reserve University).

Poster 25 - Reduction Of Physical Aging In Copper Nanoparticle Pillared CMSMs Derived From A Polyimide Precursor Incorporating The Bis(Phenyl)Fluorene-Based Cardo Moiety And Uniformly Dispersed Metal Organic Polyhedra-18 (MOP-18)

Masoumeh Tajik Asl (The University Of Texas At Dallas)*; John Ferraris (University Of Texas At Dallas); Kenneth Balkus (Univerisity Of Texas At Dallas); Inga Musselman (University Of Texas At Dallas).

Poster 26 - Forward Osmosis With Freeze Concentration For Mining And Metals Effluents

Noel Devaere (University Of Toronto)*; Runlin Yuan (University Of Toronto); Vladimiros Papangelakis (University Of Toronto).

Poster 27 - Scale-Up Of Membrane Distillation For Treating Produced Waters From Oil And Gas Industry

Ritesh Pawar (University Of Pittsburgh)*; Zhewei Zhang (University Of Pittsburgh); Radisav Vidic (Pitt).

Poster 28 - Incorporation Of Aquaporins Into 2D Materials And Membranes Towards Water Desalination

Ronald J Vogler (University Of Texas At Austin)*; Yu-Ming Tu (The University Of Texas At Austin); Benny Freeman (The University Of Texas At Austin); Manish Kumar (University Of Texas At Austin).

Poster 29 - Solute-Solute Selectivity In Membrane Separation: Definition, Measurement, And Process Scale Evaluation

Ruoyu Wang (Vanderbilt University)*; Shihong Lin (Vanderbilt University).

Poster 30 - Gas Permeation Study With SSZ-13 Zeolite Crystals And Membranes

Ninad Anjikar (Cleveland State University); Shaowei Yang (Cleveland State University)*.

Poster 31 - Fertilizer-Based Liquid Desiccants: A Novel Concept For Energy

Efficient Greenhouse Dehumidification And Water Recycling

Sarah Moussaddy (Oakland University); Jonathan Maisonneuve (Oakland University)*.

Poster 32 - Modification Of Polyamide Reverse Osmosis Membranes For The Enhanced Urea Rejection

Shahriar Habib (The University Of Alabama)*; Steven T Weinman (The University Of Alabama).

Poster 33 - High Purification Of Binary Protein Mixtures Having Close Molecular Weights By Ultrafiltration

Lixin Feng (New Jersey Institute Of Technology); Yufeng Song (New Jersey Institute Of Technology); Sagnik Basuray (New Jersey Institute Of Technology); Kamalesh K Sirkar (New Jersey Institute Of Technology)*; Solomon Isu (University Of Arkansas); Ranil Wickramasinghe (University Of Arkansas).

Poster 34 - Optimization Of The Synthesis Conditions For The Preparation Of Na^+-Gated Nanochannel Membranes For Ammonia (NH_3) Separation

Surya Padinjarekutt (Rensselaer Polytechnic Institute)*; Huazheng Li (University At Buffalo); Bratin Sengupta (University At Buffalo); Dinesh Behera (University At Buffalo); Miao Yu (University At Buffalo).

Poster 35 - New Applications Of Biomimetic Hollow Fiber Forward Osmosis Membranes

Thomas Luxbacher (Anton Paar Gmbh)*; Hermina Buksek (University Of Maribor); Irena Ban (University Of Maribor); Claus Helix-Nielsen (Technical University Of Denmark); Irena Petrinic (University Of Maribor).

Poster 36 - Effect Of Free Volume Modification On The Physical Aging Behavior Of Amine-Functionalized PIM-1 Membranes For Gas Separation

Taigyu Joo (Massachusetts Institute Of Technology)*; Katherine Mizrahi Rodriguez (MIT); Hyunhee Lee (MIT); Zachary Smith (MIT).

Poster 37 - Automated And Field-Deployable MF/UF System For Fouling Evaluation

Weiming Qi (Clemson University)*; David Ladner (Clemson).

Poster 38 - Key Factors Dominating Natural Gas Sweetening Efficiency Using Polymer Membranes

Yang Liu (Aramco Americas)*; John Yang (Aramco Americas).

Poster 39 - Zwitterionic Membrane With Phenyl Acrylate For Heavy Metal Removal From Wastewaters

Yihung Lin (Auburn University)*; Jung Min Kim (University Of Virginia); Yuyang Wang (Auburn University); Maria Auad (Auburn University); Bryan S Beckingham (Auburn University).

Poster 40 - Modeling Self-Assembly And Nonsolvent Induced Phase Separation In Block Copolymer Membranes

Anthony J Cooper (University Of California, Santa Barbara)*; Douglas Grzetic (University Of California, Santa Barbara); Alexander Bridge (The University Of Texas At Austin); Kris Delaney (UC Santa Barbara); Benny Freeman (The University Of Texas At Austin); Glenn Fredrickson (UC Santa Barbara).

Poster 41 - Microstructure Evolution During Nonsolvent-Induced Phase Separation: A Joint Experimental-Computational Investigation Into Membrane Formation

Alexander Bridge (The University Of Texas At Austin)*; Anthony J Cooper (University Of California, Santa Barbara); Jan Garcia (UC Santa Barbara); Matthew Santoso (The University Of Texas At Austin); Glenn Fredrickson (UC Santa Barbara); Benny Freeman (The University Of Texas At Austin).

Poster 42 - Understanding Mechanical Behavior Of Porous Polymeric Stamps During Large-Area Metal-Assisted Chemical Imprinting Of Silicon

Emmanuel Dasinor (Arizona State University)*; Aliaksandr Sharstniou (Arizona State University); Bruno Azaredo (Arizona State University); Yifu Ding (Membrane Science, Engineering, And Technology

(MAST) Center, Department Of Mechanical Engineering, University Of Colorado Boulder).

Poster 43 - Ion And Water Transport Mechanisms In PEGDA Membranes

Everett S Zofchak (The University Of Texas At Austin)*; Zidan Zhang (The University Of Texas At Austin); Harnoor Sachar (The University Of Texas At Austin); Nico Marioni (The University Of Texas At Austin); Sanket Kadulkar (The University Of Texas At Austin); Tyler J Duncan (The University Of Texas At Austin); Benny Freeman (The University Of Texas At Austin); Venkat Ganesan (The University Of Texas At Austin).

Poster 44 - A TiO_2-Au Janus Membrane Has Potential To Enable Self-Pumping Flow By Light

Yuhang Fang (Purdue University)*; Jefferey Moran (George Mason University); David M Warsinger (Purdue).

Poster 45 - Investigating The Effect Of Acid Strength On Process Efficiency And Product Usability In Ammonia Recovery From Wastewater Using Membrane Contactors

Abhishek Dutta (University Of British Columbia); Jongho Lee (University Of British Columbia)*.

Poster 46 - Recovery Of Medium-Chain Fatty Acids From Organic Waste Stream Using Supported Liquid Membranes

Hiroki Fukuda (University Of British Columbia); Jongho Lee (University Of British Columbia)*.

Poster 47 - Impact Of Ion-Ion Correlated Motion On Salt Transport In Solvated Ion Exchange Membranes

Nico Marioni (The University Of Texas At Austin)*; Zidan Zhang (The University Of Texas At Austin); Harnoor Sachar (The University Of Texas At Austin); Everett S Zofchak (The University Of Texas At Austin); Sanket Kadulkar (The University Of Texas At Austin); Tyler J Duncan (The University Of Texas At Austin); Benny Freeman (The University Of Texas At Austin); Venkat Ganesan (The University Of Texas At Austin).

Poster 48 - Understanding The Selectivity Between Like-Charged Ions By Modified Ion Exchange Membranes

Meng Shen (California State University, Fullerton)*; Lingchen Kong (George Washington University); Xitong Liu (George Washington University); Eric Palacios (California State University, Fullerton).

Poster 49 - Diffusion Coefficients Of Condensed Counterions In Ion-Exchange Membranes: Application Of Screening Length Scaling Relationship In Concentrated Electrolytes

Yuxuan Huang (Columbia University)*; Hanqing Fan (Columbia University); Ngai Yin Yip (Columbia University).

Poster 50 - Immobilized Molten Salt (IMS) Membrane For Ammonia Separation

Moses T Adejumo (Clarkson University)*; Simona Liguori (Clarkson University); Omid Jazani (Clarkson University).

Poster 51 - Interfacial Polymerization Kinetics: New Insights On Film Formation Using In-Situ Microscopy And Particle-Tracking

Adi Ben Zvi (Technion)*; Guy Z Ramon (Technion - Israel Institute Of Technology).

Poster 52 - UV-Grafting Of Zwitterionic Hydrogel Stabilizing ZIF-67 Nanoparticles On Polyacrylonitrile Membrane For Enhanced Antifouling In Wastewater Treatment

Akshay Modi (Ben Gurion University Of The Negev)*; Nada Sami Abdalhadi (Ben Gurion University Of The Negev); Ron Kasher (Ben Gurion University Of The Negev).

Poster 53 - Ultrathin-Film Composite Membranes Based On Perfluoropolymers For Water Desalination

Ameya Tandel (State University Of New York, Buffalo, NY)*; Haiqing Lin (The State University Of New York At Buffalo).

Poster 54 - Synthesizing DAC Hydrophobic Polymers With Various Form Factors

Ani Nazari (Arizona State University)*; Matthew Green (Arizona State University).

Poster 55 - Modification Of ZIF-8 Membranes By ZnO Deposition For Propylene/Propane Separation

Anil Ronte (Oklahoma State University)*; SHAILESH DANGWAL (Oklahoma State University); Phadindra Wagle (Oklahoma State University); Elena Echeverria (Oklahoma State University); Lin Han (University Of Akron); Jiahua Zhu (University Of Akron); Dave Mcilroy (Oklahoma State University); Seok Jhin Kim (Oklahoma State University).

Poster 56 - Upcycled Polyvinyl Chloride (PVC) Electrospun Nanofibers For Water Treatment: From Waste PVC Pipe To Membrane

ATTA UR RAZZAQ (The University Of Alabama)*; Milad Rabbani Esfahani (The University Of Alabama).

Poster 57 - Observing The Transport Behavior Of Carboxylate Salts Of Different Chain Lengths Through PEGDA-PEGMA Membranes At Varied Concentration

Antara Mazumder (Auburn University)*; Jung Min Kim (University Of Virginia); Alexandra Heist (Rollins); Bryan S Beckingham (Auburn University).

Poster 58 - Reactive Reverse Osmosis Membranes For Dissolved Silica Antifouling And Concentration Polarization Reduction By Generating Oxygen Bubbles

Weiliang Bai (University Of Texas At Austin)*; Ruizhe Xu (University Of Texas At Austin); Adam Boehnke (University Of Texas At Austin); Navid Saleh (University Of Texas At Austin); Manish Kumar (University Of Texas At Austin).

Poster 59 - Zwitterions And Their Influence Block Polymer Morphology

Bradley Grim (Arizona State University)*; Matthew Green (Arizona State University); Frederick Beyer (Army Research Laboratory).

Poster 60 - Organometallic Hybrid Nanofilms With Highly Interconnected Pore Architecture For Ultrafast Solvent Transport With Precise Molecular Separation

Bratin Sengupta (University At Buffalo)*; Qiaobei Dong (Gas Technology Institute); Ji Jiang (Rensselaer Polytechnic Institute); Dinesh Behera (University At Buffalo); Fanglei Zhou (Membrane Technology And Research); Miao Yu (University At Buffalo).

Poster 61 - Graft Polymerization Of Nanofiltration Membrane Using Oppositely Charged Monomers For The Efficient Removal Of Carbamazepine From Wastewater Effluents

Brhanu Kelali Desta (Ben-Gurion University Of The Negev)*; Akshay Modi (Ben Gurion University Of The Negev); Ron Kasher (Ben Gurion University Of The Negev).

Poster 62 - Highly Charged Ion-Exchange Membranes For Treatment Of Brines Via Electrodialysis

Carolina Espinoza (University Of Michigan)*; José Carlos Díaz (University Of Michigan); Kevin Kim (University Of Michigan); Jovan Kamcev (University Of Michigan).

Poster 63 - Melamine-Based Crosslinked Poly(Ethylene Glycol) (PEG) Membranes For Sour Gas Separations

Dana Wong (Aramco Americas)*.

Poster 64 - Integration Of Eco-Friendly Materials And Scaled-Up Fabrication Of Polymeric Membranes For Water Purification

David Lu (University Of Kentucky)*; Kwangjun Jung (Georgia Institute Of Technology); Ju Young Shim (Georgia Institute Of Technology); Alex Williams (University Of Kentucky); Kassandra Opfar (University Of Kentucky); Tequila Harris (Georgia Institute Of Technology); Isabel Escobar (University Of Kentucky).

Poster 65 - Counter-Ion Activation Energy Of Diffusion In Ion Exchange Membranes.

David Kitto (University of Michigan)*; Carolina Espinoza (University of Michigan); José C. Díaz (University of Michigan); Gregory M. Reimonn (University of Michigan); Jovan Kamcev (University of Michigan).

Poster 66 - High Flux Vapor-Gap Membranes For Reverse Osmosis Through The Inclusion Of Asymmetric Membrane Pore Structures

Elizabeth A Hjelvik (University Of Colorado Boulder)*; Kian P Lopez (University Of Colorado Boulder); Anthony Straub (University Of Colorado Boulder).

Poster 67 - Performance Of Superhydrophobic Asymmetric Electrospun Membrane In Vacuum-Air Gap Membrane Distillation

Hamid Fattahi Juybari (Amirkabir University of Technology; Purdue University)*; Mohammad Karimi (Amirkabir University of Technology) ; David M Warsinger (Purdue University).

Poster 68 - Thermal And Non-Thermal Effect And Energy Distribution In Microwave-Assisted Catalytic Membrane Filtration For Bacteriophage MS2 Inactivation

Fangzhou Liu (NJIT)*; Wen Zhang (New Jersey Institute Of Technology).

Poster 69 - Selective Separation Using Dual-Functional Nanofiltration And Adsorptive Membranes

Francisco Leniz (University Of Kentucky)*; Hailey J Craft (University Of Kentucky); Ronald Vogler (University Of Kentucky); Dibakar Bhattacharyya (U. Kentucky).

Poster 70 - Thin-Film Composite Membranes Based On Poly(Ethylene Oxide) For CO_2/N_2 Separation

Gengyi Zhang (The State University Of New York At Buffalo)*; Thien N Tran (University At Buffalo, The State University Of New York); Liang Huang (The State University Of New York At Buffalo); Liang Huang (The State University Of New York At Buffalo); Erda Deng (State University Of New York At Buffalo); Adrienne Blevins (Materials Science And Engineering Program, University Of Colorado Boulder); Wenji Guo (University At Buffalo.); Yifu Ding (Membrane Science, Engineering, And Technology (MAST) Center, Department Of Mechanical Engineering, University Of Colorado Boulder); Haiqing Lin (The State University Of New York At Buffalo).

Poster 71 - Network Nanostructured MOF To Enable Percolation And Enhanced Gas Transport

Hyunhee Lee (MIT)*; Won Seok Chi (Chonnam National University); Moonjoo Lee (Samsung); Ke Zhang (Aramco Americas); Katherine Mizrahi Rodriguez (MIT); Stephen Dewitt (MIT); Zachary Smith (MIT).

Poster 72 - Polymer Adsorber For Capturing Chemotherapy Drugs Before They Spread Through The Body

Hee Jeung Oh (Pennsylvania State University)*.

Poster 73 - Thin Film Composite Membranes With Novel Quaternary Ammonium Functionalized Poly(Arylene Ether Sulfone) Copolymers For Direct Air Capture

Hoda Shokrollahzadeh Behbahani (Arizona State University)*; Matthew Green (Arizona State University).

Poster 74 - Synthesis And Characterization Of CO_2-Selective Membranes With Poly(Ethylene Formamidine)

Jingying Hu (The Ohio State University)*; Yang Han (The Ohio State University); Winston Ho (OSU).

Poster 75 - Separation Of Trace Organics From Aqueous Streams Using Carbonaceous Membranes

Haley D White (Georgia Institute Of Technology)*; Ryan Lively (Georgia Tech).

Poster 76 - Human Skin Like Biomimetic Membranes For Breathable Protective Fabrics

Hyeonji Oh (University Of Texas At Austin)*; Yu-Ming Tu (The University of Texas at Austin); Benny Freeman (University of Texas at Austin); Manish Kumar (University of Texas at Austin).

Poster 77 - Separation Performance Of Membrane With Intrinsic Porosity For Flue Gas Purification

Petr Stanovsky (Institute Of Chemical Process Fundamentals Of The Czech Academy Of Sciences); Andrea Zitkova (Institute Of Chemical Process Fundamentals Of The Czech Academy Of Sciences); Bibiana COMESANA GANDARA (School Of Chemistry, University Of Edinburgh); Neil Mckeown (School Of Chemistry, University Of Edinburgh); Johannes Carolus (John) Jansen (CNR); Karel Friess (University Of Chemistry And Technology Prague); Jason Bara (University Of Alabama); Pavel Izak (Institute Of Chemical Process Fundamentals Of The Czech Academy Of Sciences)*.

Poster 78 - On The Order And Orientation In Liquid Crystalline Polymer Membranes For Gas Separation

Joey Kloos (Eindhoven University Of Technology)*; Nico Jansen (Eindhoven University Of Technology); Menno Houben (Eindhoven University Of Technology); Anna Casimiro (Eindhoven University Of Technology); Johan Lub (Eindhoven University Of Technology); Zandrie Borneman (Eindhoven University Of Technology); Albert Schenning (Eindhoven University Of Technology); Kitty Nijmeijer (Eijndhoven).

Poster 79 - Novel Ceramic Support Synthesis For Nanometer-Scale Thin Film Membranes Deposition

Janel Abbott (Global Research And Development, Inc.)*.

Poster 80 - Evaluation Of The Permeation Characteristics Of Pd_{82}-Ag_{15}-Y_3 Membrane For Hydrogen Separation At Different Temperature, Pressure And In Presence Of Other Gases

Omid Jazani (Clarkson University)*; Moses T Adejumo (Clarkson University); Simona Liguori (Clarkson University).

Poster 81 - Ionic Conductivity Of Ion-Exchange Membranes: Measurement Techniques, Salt Concentration Dependence, And Best Practices.

José Carlos Díaz (University Of Michigan)*; David Kitto (University Of Michigan - Kamcev Lab); Jovan Kamcev (University Of Michigan).

Poster 82 - High Pressure Reverse Osmosis Membrane Modeling And Characterization

Jeffrey R Mccutcheon (University Of Connecticut)*; Yara Suleiman (University Of Connecticut); Danh Nguyen (University Of Connecticut); Mi Zhang (University Of Connecticut); Ramesh Bhave (Ornl); Ramanan Sankaran (Oak Ridge National Laboratory); Luke Malencica (Oak Ridge National Laboratory); Kyle Cole (Michigan State University); Mayur Ostwal (University Of Connecticut); Sina Shahbazmohamadi (University Of Connecticut); Menachem Elimelech (Yale University); Ying Li (University Of Connecticut); Marcus Day (National Renewable Energy Laboratory).

Poster 83 - Rubbery Polymer-Based Membranes For NGL Recovery From Natural Gas

Yang Liu (Aramco Americas)*; John Yang (Aramco Americas); Milind Vaidya (Saudi Aramco); Sebastien Duval (Saudi Aramco); Feras Hamad (Saudi Aramco).

Poster 84 - Preparation Of Polyethersulfone Membranes On Non-Woven Fabric Supports

Joseph T Danner (The University Of Alabama)*; Blake Jarrell (The University Of Alabama); Steven T Weinman (The University Of Alabama).

Poster 85 - Influence Of Novel Processing Techniques On Poly(Benzimidazole) Gas Separation Membranes

Julian Richardson (University Of Texas, Austin)*; Benny Freeman (The University Of Texas At Austin).

Poster 86 - Lithium Extraction And Purification From Geothermal Brines Via Selective Bipolar Electrodialysis

Aubrey Quigley (University Of Texas At Austin); Kevin Reimund (University Of Texas At Austin)*; Benjamin Pedretti (University Of Texas At Austin); Kristofer Gleason (University Of Texas At Austin); Nathaniel Lynd (The University Of Texas At Austin); Benny Freeman (The University Of Texas At Austin).

Poster 87 - Lithium-Magnesium Selectivity In Glassy Cellulose Acetate Membranes)

Kevin Reimund (University Of Texas At Austin)*; Paul Irving (University Of Texas At Austin); Venkat Ganesan (University Of Texas At Austin); Benny Freeman (The University Of Texas At Austin).

Poster 88 - High Performance Polyamide-Polyarylate TFC Membranes Via Selective Etching Of Amide/Arylate Bonds

Chenhao Yao (University Of Texas At Austin)*; Manish Kumar (University Of Texas At Austin).

Poster 90 - Fouling Resistant And Tunable Polyampholyte Selective Layers

Luca Mazzaferro (Tufts University)*; Samuel J Lounder (Tufts); Ayse Asatekin (Tufts).

Poster 91 - Enhancing The Salt Rejections Of Covalent Organic Framework Membranes By Creating Dual-Layer Membrane Structure

Miguel Jaimes (UWM)*; Zhiqin Qiang (University Of Wisconsin-Milwaukee); Rahul Khandge (University Of Wisconsin - Milwaukee); Ho Kuan Yu (University Of Wisconsin - Milwaukee); Weiling Xia (University Of Wisconsin - Milwaukee); Rachel Clark (University Of Wisconsin - Milwaukee); Xiaoli Ma (University Of Wisconsin-Milwaukee).

Poster 92 - Ethylene And Ethane Transport Properties Of Hydrogen-Stable Ag+-Containing Facilitated Transport Membranes

Matthew N Davenport (The University Of Texas At Austin)*; Caitlin Bentley (The University Of Texas At Austin); Joan Brennecke (The University Of Texas At Austin); Benny Freeman (The University Of Texas At Austin).

Poster 93 - Characterizing Sulfonated Pentablock Terpolymer Nexar™ As Pervaporation Membranes For Treatment Of Inland Desalination Concentrates

Mariana Hernandez Molina (Arizona State University)*.

Poster 94 - MOF-Functionalized Thin-Film Nanocomposite Membranes For Enhanced PFAS Removal From Water

Medha Kasula (The University Of Alabama)*; Milad Rabbani Esfahani (The University Of Alabama); Tin Le (The University Of Alabama).

Poster 95 - Pentiptycene-Containing Polybenzimidazole-Based Semi-Interpenetrating Network (S-IPN) Membranes For H_2/CO_2 Separation

Mengdi Liu (University Of Notre Dame)*.

Poster 96 - Defect-Free Matrimid® Asymmetric Gas Separation Membranes Prepared With Dihydrolevoglucosenone (Cyrenetm) As A Greener Polar Aprotic Solvent

Noah P Wamble (University Of Texas At Austin)*; Alexander Bridge (The University Of Texas At Austin); Joan Brennecke (The University Of Texas At Austin); Benny Freeman (The University Of Texas At Austin).

Poster 97 - Preparation Of Highly Permeable Substrates By Vapor-Induced Phase Separation For CO_2-Selective Composite Membranes

Ruizhi Pang (The Ohio State University)*; Yutong Yang (The Ohio State University); Yang Han (The Ohio State University); Kai Chen (The Ohio State University); Winston Ho (OSU).

Poster 98 - Water And Vapor Transport Through Angstrom-Scale Pores In Atomically Thin Graphene Membranes

Piran Kidambi (Vanderbilt University)*.

Poster 99 - Blending Of Multi-Functionalized Multi-Walled Carbon Nanotubes Improves Antifouling And Antibacterial Properties Of Polyethersulfone Hollow Fiber Membranes

PREETY KUMARI (Indian Institute Of Technology Bombay)*; Jayesh Bellare (Indian Institute Of Technology Bombay).

Poster 100 - So, You Think You Can Graft? Avoiding Pitfalls In Characterization When

"Grafting From" Polyethersulfone Membranes

Priyanka Suresh (Case Western Reserve University)*; Megan Sibley (Case Western Reserve University); Amy Che (Case Western Reserve University); Lauren M Ward (The University Of Alabama); Steven T Weinman (The University Of Alabama); Christine E Duval (Case Western Reserve University).

Poster 101 - Transport And Co-Transport Of Methanol And Carboxylate In Sulfonated Cation Exchange Membranes With Poly(Ethylene Glycol) Phenyl Ether Acrylate As A Blocking Group

Pravin Parasakthi Aravindhan (Auburn University)*; Jung Min Kim (University Of Virginia); Bryan S Beckingham (Auburn University).

Poster 102 - Two-Step Thermoresponsive Ultrafiltration Membrane Via Polymerization Of Lyotropic Liquid Crystals

Younes Saadat (University Of Oklahoma); Reza Foudazi (University Of Oklahoma)*.

Poster 103 - Pressure Induced Diffusion In Hydrated Membranes

Rahul Sujanani (The University Of Texas At Austin)*; Kevin Reimund (University Of Texas At Austin); Kris Gleason (The University Of Texas At Austin); Benny Freeman (The University Of Texas At Austin).

Poster 104 - Alcohol-Carboxylate Transport In Phenyl-Based Membranes With Varying Side Chain Length

Sarah A Gaston (Auburn University)*; Jung Min Kim (University Of Virginia); Bryan S Beckingham (Auburn University); Minjae Jung (Auburn University).

Poster 105 - Cross-Linkable/ Self-Assembling Terpolymer Membranes For Brackish Water Desalination

Samuel J Lounder (Tufts)*; Ayse Asatekin (Tufts).

Poster 106 - Fabrication Of Patterned Microporous Membranes With Enhanced Pore Connectivity And High Salt-Scaling Resistance Using Confined Thermally Induced Phase Separation Under Micromolding

Shouhong Fan (University Of Colorado At Boulder)*; John Chau (New Jersey Institute Of Technology); Adrienne Blevins (Materials Science And Engineering Program, University Of Colorado Boulder); Duong T. Nguyen (University Of Colorado Boulder); Jaylene Martinez (CU Boulder Mechanical Engineering); Anthony Straub (University Of Colorado Boulder); Kamalesh K Sirkar (New Jersey Institute Of Technology); Yifu Ding (Membrane Science, Engineering, And Technology (MAST) Center, Department Of Mechanical Engineering, University Of Colorado Boulder).

Poster 107 - Salt Transport In Sulfonated Polysulfone Membranes Exposed To High Salinity Conditions

Sean M Bannon (University Of Virginia)*; Geoffrey M Geise (University Of Virginia).

Poster 108 - Ionic MOF-Cellulose Acetate Electrospun Membranes For Selective Removal Of Selenium From Water

Sweta A Modak (The University Of Alabama)*; Milad Rabbani Esfahani (The University Of Alabama); ATTA UR RAZZAQ (The University Of Alabama).

Poster 109 - Remarkable Effect Of Low Loading Of Metal-Organic Polyhedra On CO_2/N_2 Separation Properties Of Cross-Linked Polyethers

Taliehsadat Alebrahim (The State University Of New York At Buffalo)*; Liang Huang (The State University Of New York At Buffalo); Heshali Welgama (The State University Of New York At Buffalo); Shiwang Cheng (Michigan State University); Durga Acharya (CSIRO); Cara Doherty (CSIRO); Anita Hill (CSIRO); Clayton Rumsey (The State University Of New York At Buffalo); Martin Trebbin (The State University Of New York At Buffalo); Tim Cook (University At Buffalo); Haiqing Lin (The State University Of New York At Buffalo).

Poster 110 - High-Permeance Organic Solvent Nanofiltration Composite Membranes Synthesised With Kevlar Aramid Nanofiber And Graphene Oxide And Post-Treated With Calcium Chloride

Tanya Srivastava (Texas Tech University)*; I-Min Hsieh (Texas Tech University); Mahdi Malmali (Texas Tech University).

Poster 111 - Synthesis, Additive Manufacturing, And Performance Of Polysulfone Membranes For Separation Processes

Taysha B Telenar (Arizona State University)*.

Poster 112 - Membrane Adsorbents Comprising Self-Assembled Inorganic Nanocages (SINCs) For Direct Air Capture

Thien N Tran (University At Buffalo, The State University Of New York)*; Alisa Chakraborty (The State University Of New York At Buffalo); Shweta Singh (University At Buffalo, The State University Of New York); Amanda Grass (University At Buffalo, The State University Of New York); Tim Cook (University At Buffalo); Haiqing Lin (The State University Of New York At Buffalo).

Poster 113 - Facile Fabrication Of Organosilica Membrane For High-Temperature H_2/CO_2 Separation Using Oxygen Plasma Treatment Of Polysiloxane

Vinh Bui (University At Buffalo)*; Haiqing Lin (The State University Of New York At Buffalo).

Poster 114 - Polyethyleneimine-Based Surface Treatment For Facile Synthesis Of Conductive Membranes: Active Material Doesn't Matter

Wei Zhang (University Of North Carolina At Chapel Hill)*; Orlando Coronell (University Of North Carolina At Chapel Hill).

Poster 115 - Nanophotonic Enhanced Pervaporation Desalination Membranes

Yusi Li (Arizona State University)*.

Poster 116 - Structure-Transport Studies Of Carbon Molecular Sieve Membranes For Wastewater Treatment

Young Hee Yoon (Georgia Institute Of Technology)*; Daniel O'Nolan (Georgia Institute Of Technology); Michelle Beauvais (Stony Brook University); Karena Chapman (Stony Brook University); Ryan Lively (Georgia Tech).

Poster 117 - Crosslinked Triptycene-Containing Polybenzoxazole Membranes With Model Network Structures With High Performance Gas Separations

Ziwei Dai (University Of Notre Dame)*.

Poster 118 - Synthesis Of ZIF Membranes By A Vapor-Phase Seeding Method For Olefin/Paraffin Separation

Zhiqin Qiang (University Of Wisconsin-Milwaukee)*; Weiling Xia (University Of Wisconsin-Milwaukee); Rachel Clark (University Of Wisconsin-Milwaukee); Zihao Yi (University Of Wisconsin-Milwaukee); Rahul Sampat Khandge (University Of Wisconsin-Milwaukee); Xiaoli Ma (University Of Wisconsin-Milwaukee).

Poster 119 - Direct Ocean Capture Using Membrane Contactors

Austin R Lieber (University Of Pittsburgh)*; Joanna Rivero (University Of Pittsburgh); Donna Hildebrandt (University Of Pittsburgh); Chris Snodgrass (University Of Pittsburgh); William Gamble (University Of Pittsburgh); Zoe Neal (University Of Pittsburgh); Shanna Davidson (University Of Pittsburgh); Huda Usman (University Of Pittsburgh); Tagbo Niepa (University Of Pittsburgh); Katherine Hornbostel (University Of Pittsburgh).

Poster 120 - Polydopamine Enhances Longevity Of Copper (II) Oxide Nanoparticles On Polypropylene Spacers And Enhances Biofouling Control In Reverse Osmosis

Amos Taiswa (Montana Tech)*; Jesicca Andriollo (Montana Tech); Katherine Zodrow (Montan Tech); Jack Skinner (Montana Tech).

Poster 121 - System-Level Approach For Model Optimization And Testing Of Low-Energy Inland Brackish Water Desalination In Arid Climates

Arianna Q Tariqi (University Of Arizona)*; Kerri Hickenbottom (University Of Arizona); Vicky Karanikola (University Of Arizona); Anthony Straub

(University Of Colorado Boulder); Luis Cruzado (Bureau Of Reclamation).

Poster 122 - Membrane Technologies As Strategies For Spaceflight Wastewater Management

Elisabeth R Thomas (Arizona State University)*; Mary Laura Lind (Arizona State University).

Poster 123 - Synthesis And Performance Characterization Of Zwitterionic Poly(Arylene Ether Sulfone) Copolymers

Husain Mithaiwala (Arizona State University)*; Matthew Green (Arizona State University); Seok-Jun Cho (Watts Water Technology); Dorian Cauceglia (Watts Water Technology); Daniel Galperin (Watts Water Technology).

Poster 124 - Superhydrophobic Laser-Induced Graphene Polyethersulfone Membrane For Desalination Application With Vacuum Membrane Distillation

Md Hasib Al Mahbub (Texas Tech University)*; Amit Kumar Thakur (Texas Tech University); Tanya Srivastava (Texas Tech University); I-Min Hsieh (Texas Tech University); Mahdi Malmali (Texas Tech University).

Poster 125 - Unraveling The Scaling Mechanism In Membrane Distillation For The Treatment Of Hypersaline Brine

I-Min Hsieh (Texas Tech University)*; Bosong Lin (Texas Tech University); Hadis Matinpour (University Of California, Santa Barbara); Mahdi Malmali (Texas Tech University).

Poster 126 - Electrocatalytic Upcycling Of Nitrate Wastewater To Ammonia Fertilizer Via Electrified Membrane)

Jianan Gao (New Jersey Institute Of Technology)*; Yifan Li (Qingdao University Of Technology); Bo Jiang (Qingdao University Of Technology); Taha Marhaba (New Jersey Institute Of Technology); Wen Zhang (New Jersey Institute Of Technology).

Poster 127 - The Effect Of Channel Height Variation In A Flat Sheet Membrane: A CO_2 Separation Case Study

Joanna Rivero (University Of Pittsburgh)*; Marcos Da Conceicao (University Of Toledo); Leo Nemetz (University Of Toledo); Glenn Lipscomb (University Of Toledo, Toledo); Katherine Hornbostel (University Of Pittsburgh).

Poster 128 - Optimizing Of Polyamide Nanofiltration Membrane For Lithium Enrichment From Salt Lake Brine

Juan Zhai (Texas Tech University)*; Sunday P Adu (Texas Tech University); Yuexiao Shen (Texas Tech University).

Poster 129 - Gypsum Scaling In Membrane Distillation: Impacts Of Temperature And Vapor Flux

Kofi SS Christie (Princeton University)*; Thomas A Horseman (Vanderbilt University); Ruoyu Wang (Vanderbilt University); Chunlei Su (Chinese Academy Of Sciences); Shihong Lin (Vanderbilt University).

Poster 130 - Copper Recovery With Iminodiacetate-Functionalized Cation Exchange Membranes

Lauren A Mazurowski (Yale University)*; Ryan Duchanois (Yale University); Menachem Elimelech (Yale University).

Poster 131 - Membrane Fouling By Mixtures Of Oppositely Charged Particles

Huang Teik Lay (NTU SMTC)*; Rong Wang (Nanyang Technological University); Jia Wei Chew (Nanyang).

Poster 132 - Investigation Of Flow Dynamics In A Spacer-Filled Membrane Distillation Channel

Bosong Lin (Texas Tech University)*; Hadis Matinpour (University Of California, Santa Barbara); Mahdi Malmali (Texas Tech University).

Poster 133 - Evaluating Commercial Membranes For The Desalination Of Produce Water Using Direct Contact Membrane Distillation

Lucy M Camacho (Texas A&M University-Kingsville, Department Of Environmental Engineering)*;

Oluwadamilola T. Otoki (Texas A&M University-Kingsville).

Ryan Duchanois (Yale University)*; Menachem Elimelech (Yale University).

Poster 134 - The Effect Of Feed Water Temperature And Permeate Flux On The Scaling Behavior Of Membrane Distillation Systems

Marion Bellier (Arizona State University)*; Francois Perreault (Arizona State University); Jin Wen Soh (Arizona State University); Qilin Li (Rice University); Mohamed EA Ali (Desert Research Center); Mustafa Aboelfadel (Desert Research Center).

Poster 135 - Permeation Through Pore Defects In Pilot-Scale Membrane Distillation

Mukta Hardikar (The University Of Arizona)*; Luisa Ikner (University Of Arizona); Kerri Hickenbottom (U. Arizona); Andrea Achilli (The University Of Arizona).

Poster 136 - Thermal Storage Management Integrated With Solar Driven Membrane Distillation For Process Intensification

Mikah M Inkawhich (University Of Arizona)*; Jeb C Shingler (University Of Arizona); Spencer Ketchum (University Of Arizona); Wei Pan (DWP Energy Solutions); Robert Norwood (University Of Arizona); Andrea Achilli (The University Of Arizona); Kerri Hickenbottom (U. Arizona).

Poster 137 - Organic Solvent Nanofiltration Using Covalent Organic Frameworks And Unravelling The Effect Of Solvent-Membrane Interactions On Permeate Flux

Parisa Taheri (University Of Houston)*; Devin Shaffer (University Of Houston).

Poster 138 - Development Of 3D-Printed Membranes For The Production Of Low-Carbon Intensity Biofuels

Rebecca Lee (University Of Connecticut)*; Jeffrey R Mccutcheon (University Of Connecticut); Mayur Ostwal (University Of Connecticut).

Poster 139 - Designing Polymeric Membranes With Coordination Chemistry For High-Precision Ion Separations

Poster 140 - Interdigitated Electrodes On Polymeric Membrane Surfaces To Control Biofouling

Soobin Cho (Rice University)*; Bezawit Getachew (Rice University).

Poster 141 - Ozone-Membrane Disitillation And Conventional Concentrate Management: A Techno-Economic Comparison

Varinia V Felix (The University Of Arizona)*; Kerri Hickenbottom (The University Of Arizona); Andrea Achilli (The University Of Arizona); Mukta Hardikar (The University Of Arizona).

Poster 142 - Membrane Nutrient Recovery From Poultry Wastewater

Zahra Anari (Penn State University)*; Lauren Greenlee (Penn State University).

Poster 143 - Transport And Physiochemical Properties Of Polyether-Based Membranes As Alternatives To Nafion For Nitrogen Reduction Fuel Cells

Brock J Hunter (Auburn University)*; Bryan S Beckingham (Auburn University).

Poster 144 - Gas Separation Properties Of Polydimethylsiloxane-Polyphenylmethylsiloxane Copolymer Membranes

Elizabeth Haddad (University At Buffalo)*; Vinh Bui (University At Buffalo); Haiqing Lin (The State University Of New York At Buffalo).

Poster 145 - Anion Exchange Membranes For Artificial Photosynthesis

Erik E Jackson (Lawrence Berkeley National Laboratory)*; Daniel J Miller (Lawrence Berkeley National Laboratory).

Poster 146 - A Balanced Charged Polyelectrolyte Complex Nanofiltration Membranes For Effluents Desalination

Hao Huang (Ben Gurion University Of The Negev)*; Roy Bernstein (Ben Gurion University).

Poster 147 - Electrospun Styrenic Membranes With Tailorable Surface Chemistry

Joelle V Scott (Case Western Reserve University)*; Maura Sepesy (Case Western Reserve University); Tuli Banik (Case Western Reserve University); Alec Johnson (Case Western Reserve University); Christine E Duval (Case Western Reserve University).

Poster 148 - Separation Of Organic Acids For Direct Catalysis From Sugars

Katelyn Robinson (University Of Arkansas).

Poster 149 - Effects Of Applied Potential On The Performance Of The As Synthesized PSF/NMP/S-PEEK Membranes For Application In Water Remediation

Laura Brady (University Of Kentucky)*; Kate Mclarney (University Of Kentucky); Abelline K Fionah (University Of Kentucky); Isabel Escobar (University Of Kentucky).

Poster 150 - Integration Of Enzyme With Commercial Water-Filtration Membranes For Coronavirus-Sized Aerosol Capture And Protein Deactivation

Matthew D Bernard (University Of Kentucky)*; Rollie Mills (U. Kentucky); Ronald Vogler (University Of Kentucky); Jacob Concolino (University Of Kentucky); Yinai Wei (University Of Kentucky); Jeffrey Hastings (University Of Kentucky); Thomas Dziubla (University Of Kentucky); Kevin Baldridge (University Of Kentucky); Dibakar Bhattacharyya (U. Kentucky).

Poster 151 - Modification Of Thin-Film Composite Reverse Osmosis Membranes With Linear Diamines For The Improved Rejection Of Urea

Madison A Wilkins (The University Of Alabama)*; Shahriar Habib (The University Of Alabama); Steven T Weinman (The University Of Alabama).

Poster 152 - Fabrication Of Polyelectrolyte Multilayer Films Using An Unconventional Direct Ink-Writing Technique

Shawn Stephens (West Virginia University)*; Piash Kmprottoy Shariar (West Virginia University); Michael Ferrebee (West Virginia University); Rebecca Erwin (West Virginia University); Guy Cordonier (West Virginia University); Oishi Sanyal (West Virginia University); Kostas Sierros (West Virginia University).

Poster 153 - Modified Membrane Feed Spacer Mesh As Solid-Phase Antimicrobials In Food Processing

Ian Brown (The University Of Alabama); James Segars (The University Of Alabama); Robert Scott (The University Of Alabama); Ryan Summers (The University Of Alabama); Stephen Ritchie (The University Of Alabama)*.

Poster 154 - Effect Of Interfacial Polymerization Conditions On The Structure And Property Of Covalent Organic Framework Membranes

Weiling Xia (University Of Wisconsin-Milwaukee).

Poster 155 - Integrally Skinned Polybenzimidazole-derived Asymmetric Carbon Molecular Sieve Hollow Fiber Membranes for Blue Hydrogen Production

Jonggeun Seong (Los Alamos National Laboratory)*; Jeremy Lewis (Los Alamos National Laboratory), Ulises Martinez (Los Alamos National Laboratory), Harshul V. Thakkar (Los Alamos National Laboratory), John Matteson (Los Alamos National Laboratory), Kathryn A. Berchtold (Los Alamos National Laboratory), Rajinder P. Singh (Los Alamos National Laboratory).

Poster 156 - Development Of Novel Anion-Exchange Membranes For The Separation Of Full And Empty AAV Capsids

Xiaolei Hao (University Of Arkansas); Ronny Hao (University Of Arkansas); Ranil Wickramasinghe (University Of Arkansas); Xianghong Qian (University Of Arkansas)*.

Poster 157 - Towards A Universal Framework For Evaluating Mass Transport In Pressure,

Concentration, And Temperature Driven Membrane-Based Desalination Systems

Kian P Lopez (University Of Colorado Boulder)*; Ruoyu Wang (Vanderbilt University); Elizabeth A Hjelvik (University Of Colorado Boulder); Shihong Lin (Vanderbilt University); Anthony Straub (University Of Colorado Boulder).

Poster 158 - Influence of Hydroxyl Groups on Water Transport in Highly Charged IEMs

Kevin Kim (University of Michigan)

Poster 159 - A Novel Urine Flowmeter for Hospitalized In-Patients

Piyush Hota*, Bo Fu, Manni Mo, Adithya Shyamala Pandian, Sandra Miranda, Phuoc H.H. Duong, Peter Wiktor, Leslie Thomas, Erica Forzani, Mary Laura Lind.

Poster 160 - Smart-Controlled Reversible Colorimetric Sensing System

Manni Mo*, Bo Fu, Piyush Hota, Pinar Cay-Durgun, Peter Wiktor, Leslie Thomas, Mary Laura Lind, Erica Forzani.

Poster 161 - Dynamic Light Intensity Attenuation Measurement Solution Using Custom Embedded System

Bo Fu*, Manni Mo, Piyush HOta, Pinar Cay-Durgun, Peter Wiktor, Leslie Thomas, Mary Laura Lind, Erica Forzani.

Poster 162 - CO_2 Recovery In A Pilot-Scale Field Testing Post-Combustion Process Using GENERON's Hollow Fiber Membranes

Nathanael Wasson (GENERON), Ryan Collins(GENERON), Zamidi Ahmad (GENERON)*.

Poster 163 - GENERON's Hollow Fiber Membrane Durability Against Volatile Organic Compound (VOC) Contaminants: A Pilot Study.

Nathanael Wasson (GENERON), Ryan Collins(GENERON), Zamidi Ahmad (GENERON)*.

Poster 164 - Effect Of Packing Nonuniformity On The Performance Of Hollow Fiber Membrane Gas Separation Modules Fabricated From Fiber Tows

Lili Sun (University Of Toledo)*; Grigorios Panagakos (Carnegie Mellon University); Glenn Lipscomb (University Of Toledo, Toledo).

Abstracts – Oral Presentations, Monday

Presenting authors are indicated by an asterisk*. See the most up-to-date version of the program book with full abstracts at **www.membranes.org/nams~2022/**

Oral 1 – ORGANIC SOLVENT SEPARATIONS I

9:30 am (Room Palm AD, Monday)

Oral 1 - From renewable building blocks to biodegradable polymer membranes with solvent-resistant properties

Gyorgy Szekely (King Abdullah University of Science and Technology)*

To address increasing environmental concerns, green and sustainable materials that promise to replace fossil-based separation materials, and reduce waste generation are attracting considerable research attention. On one hand, the monomers need to be reactive to form a highly crosslinked thin film. On the other hand, one of the monomers need to be soluble in polar solvents, while the other monomer need to be soluble in a non-polar solvent, which is immiscible with the selected polar solvent. Our group has been exploring various green sources to fabricate solvent-resistant nanofiltration membranes. Sustainable thin-film composite (TFC) membranes were designed and fabricated via interfacial polymerization of green building blocks. Chitosan from shrimp-farming waste was derived and turned into a TFC membrane with the plant-based 2,5-furandicarboxaldehyde crosslinker. Moreover, a novel hydrophobic TFC membrane was prepared via interfacial polymerization of plant-based tannic acid and priamine monomers. The cyclohexane, benzene and aliphatic moieties of priamine are suitable to increase the hydrophobicity of the membrane surface, which in turn can result in high permeance of non-polar solvents. In addition, tannic acid is a representative polyphenol compound, which can be extracted from a diverse range of natural materials such as trees, plants, nuts and fruits. We solubilized date seed biomass (abundantly available from the multimillion-metric-ton date industry) using ionic liquids and dimethyl sulfoxide to fabricate biodegradable nanofiltration membranes. The resultant membranes were coated with mussel-inspired polydopamine via a layer-by-layer deposition method. The obtained membranes demonstrated excellent performance for solvent and oil-in-water separations. We found that the separation performance of the membranes obtained from solely green sources was competitive to the fossil-based tailored and commercial membranes.

10:00 am (Room Palm AD, Monday)

Oral 2 - Towards wired-lab organic solvent nanofiltration: is machine learning viable?

Gergo Ignacz (King Abdullah University of Science and Technology)*; Gyorgy Szekely (King Abdullah University of Science and Technology)

Organic solvent nanofiltration (OSN) is a potentially green and sustainable membrane technology. OSN has a promising future in the pharmaceutical and fine chemical industry for solvent and catalyst recovery. However, despite the many advantages of OSN, this industrial implementation is hindered by small, specialized datasets and the limited availability of predictive models. In our presentation, first, we will critically assess current challenges in OSN. We will define chemical diversity and its importance in organic solvent nanofiltration. We propose that a large, chemically diverse dataset would open opportunities for implementing predictive models utilizing the structural information of the membranes, solvent, and dissolved solutes. We present a medium-throughput system (MTS) to measure the rejection of small chemicals in OSN. Using this MTS method, we establish the OSN Database, the online and open-access dataset for OSN. The OSN Database currently contains the rejection of more than 500 molecules. We also show the correlations of rejection on different molecular and process parameters. We define the term "wired-lab" for membrane science and how chemical structure-based predictive models are the keys to automated membrane processes. We show that current predictive models implement chemical structures neither for the membrane material, solvent, or dissolved solute. We present the first structure-based prediction models for OSN using different machine learning techniques. Today, our models have the lowest root mean squared error and highest fitting scores for predicting solute rejection in OSN. Feature extraction and variable importance scoring were used to interpret the relationship across molecules in the dataset. For the first time, we visualized the effect of functional groups on rejection leveraging deep learning models. Our models are open access and available on the www.osndatabase.com website.

10:30 am (Room Palm AD, Monday)

Oral 3 - Tuning polyimide thin film composite membranes for organic solvent reverse osmosis separations via Boc protected amine solid-state crosslinking

Yacine Feliachi (Georgia Tech)*; M.G. Finn (Georgia Institute of Technology); Ryan Lively (Georgia Tech)

Organic solvent separations are energy intensive yet ubiquitous in the chemical, petrochemical, and pharmaceutical industries. Membrane technology can reduce the energy, carbon, and water intensity required in these separations significantly. However, these molecular separations often involve highly similar molecules, which reduces the efficacy of the membrane-based process. Some polymers used in organic solvent reverse osmosis (OSRO) have been known to suffer from plasticization, making them inefficient for these molecular separations. One method to avoid such effects is crosslinking the polymer. Here, we develop a method to uniformly crosslink a selective layer for a thin film composite (TFC) membrane to improve stability in organic solvents and tune separation performance. Traditional crosslinking methods are hindered by the diffusion of crosslinking agents into the polymer matrix, and this effect becomes more pronounced as crosslinking agents become bulkier. We have developed a method to introduce larger crosslinking agents that impart additional free volume to the membrane while overcoming diffusional limitations during the crosslinking infusion step. We utilize the commercial polymer Matrimid® 5218 and three crosslinking agents: para-xylylene diamine, 4,4'-biphenyldicarbontrile, and triptycene. We introduce a Boc protection method to the amines before dissolving into the membrane casting solution. After casting, the amines within the membrane are thermally deprotected to induce a solid-state crosslinking reaction. The resulting membranes were characterized by FT-IR, gel fraction, helium pycnometry, and TGA. The separations performance was tested in a crossflow OSRO permeation system for a toluene/triisopropyl benzene system and a multicomponent organic solvent mixture. The crosslinked TFC membranes showed excellent stability in organic solvents and tunable separation characteristics based on crosslinking agent and concentration of the crosslinking agent.

11:00 am (Room Palm AD, Monday)

Oral 4 - Organic Solvent Stable Graft Modified Polybenzimidazole Membranes with Customisable Rejection Profiles

Adam E Oxley (Imperial College London)*; Piers Gaffney (Queen Mary University of London); Andrew Livingston (Queen Mary University of London)

Crosslinked polybenzimidazole (PBI) membranes have been shown to have excellent stability in a range of organic solvents. However, tuning the selectivity of these membranes has remained a significant difficulty. In this work, crosslinked PBI membranes were graft modified to overcome this challenge. A flexible long chain polyether monoamine was used as the primary grafting agent. The extent of grafting was then controlled by using a competitive reaction with a small molecule monoamine, which caps a known percentage of the available grafting sites. By controlling the extent of grafting, the molecular weight cut off can be customised between 2,000 and 20,000 g mol-1, subject to the performance requirements of the membrane. This was shown to be due to the grafting of the polyether inside the permeation pathways of the membrane, and not just grafting on the top surface of the membrane.

The casting and post casting reactions were scaled up in order to manufacture 5 m batches of membrane. This membrane was rolled into 1.8" by 12" spiral wound modules, which have consistent performance to the flat sheet membranes. These membranes are currently being used for the synthesis of oligonucleotide therapeutics, peptides and sequence defined synthetic polymers.

11:30 am (Room Palm AD, Monday)

Oral 5 - Precision molecular sieving with MOF and COF membranes in organic solvents

Rifan Hardian (King Abdullah University of Science and Technology)*; Gyorgy Szekely (King Abdullah University of Science and Technology)

To meet the growing demand for precise molecular sieving in organic solvent nanofiltration, it is essential to rationally tune the pore size. In quest to achieve highly selective membranes, we investigated mixed matrix membranes (MMMs) containing metal–organic framework (MOF) fillers, owing to the regularity and ordered pore arrangement in the MOF structure [1]. The poor interfacial adhesion between the polymer matrix and the MOF filler was overcome by covalently grafting the MOF with poly(N-isopropylacrylamide) (PNIPAM) chains, which could then become entangled with the membranes' polymer matrix. This strategy was inspired by the natural phenomenon of strong interfacial adhesion between the calcified cementum of the tooth and the surrounding tissues provided by intermingled collagen fibers. We prepared a series of novel membranes comprising MOF nanoparticles with similar chemical and morphological properties but having various pore sizes (UiO-66–68-NH_2). The OSN test revealed the increase in the flux and molecular weight cut-off values with increasing MOF aperture size, which confirm the key role that MOF pores play in the filtration. Similarly, we investigated the effect of structural pore size on the molecular sieving performance of self-standing COF membranes. Four COF membranes with superior chemical stability were fabricated. The 2D structure of COF is particularly interesting to enable the formation of thin films. The coupling of OSN measurements with transport modeling supported the facts that COF structural pore size governs the filtration performance of the membranes. Employing transport modeling not only enriched the interpretation of the experimental results, but also enabled us to gain more insight into the effect of materials structure and operational parameters on the OSN performance.

[1] J. Mater. Chem. A, 2021, 9, 23793–23801.

12:00 pm (Room Palm AD, Monday)

Oral 6 - Engineering organic solvent reverse osmosis performance of vapor phase infiltrated organic-inorganic hybrid membranes

Yi Ren (Georgia Institute of Technology)*; Emily McGuinness (Georgia Institute of Technology); Benjamin Jean (Georgia Institute of Technology); Chaofan Huang (Georgia Institute of Technology); Roshan Joseph (Georgia Institute of Technology); Mark Losego (Georgia Institute of Technology); Ryan Lively (Georgia Tech)

Membrane based organic solvent separations serve as a low-energy alternative to traditional thermal-based separations. While inorganic membranes have high chemical stability to withstand the harsh chemical environments, the high cost and defect rates at scale are obstacles to commercialization. On the other hand, polymer membranes have high scalability with reasonable cost, but the low chemical stability will result in membrane swelling, plasticizing, or even complete dissolution in organic solvent environments. Vapor-phase infiltration (VPI) has emerged as a scalable, solvent-free post-treatment process that can transform polymer products into organic-inorganic hybrid materials with enhanced properties, including chemical stability which is desired for membrane based organic solvent separations. The polymeric membrane material is exposed to metal-organic vapor, which will allow infiltration of the vapor phase metal-organic molecules into the polymer. The metal-organics will grow into metal-oxide networks that homogeneously mix with polymer chains on a molecular level, prohibiting the polymer chains from large solvent-induced motions, thus hybridizing the entire material into organic-inorganic membrane with higher solvent stability. In this work, the transport of metal-organic vapors into polymer materials will be investigated to better describe and predict the reaction-diffusion transport phenomenon occurring during a typical VPI process for hybridizing polymer membranes. Then organic solvent reverse osmosis transport of the hybridized membrane will be engineered by controlling the metal loading during the VPI process.

Oral 2 – CARBON CAPTURE I

9:30 am (Room Palm CF, Monday)

Oral 7 - Thin Film Composite Membranes with Novel Quaternary Ammonium Functionalized Poly(Arylene Ether Sulfone) Copolymers For Direct Air Capture

Hoda Shokrollahzadeh Behbahani (Arizona State University)*; Matthew Green (Arizona State University)

The drastic rise in CO_2 levels of the atmosphere has amplified the natural greenhouse effect and resulted in significant global climate change. Direct air capture (DAC), i.e., capturing CO_2 directly from air is a promising approach to reduce, sequester, and reuse the atmospheric CO_2. This study proposes a novel quaternary ammonium coated thin film composite membrane (TFC) to directly capture and release CO_2 from air by the simple and energy efficient moisture swing process (MSS). The TFC utilizes quaternary ammonium-functionalized poly(arylene ether sulfone) (PAES-co-QAPAES) copolymer on porous polyethersulfone (PES) membrane acting as support to capture and release atmospheric levels of CO_2. Allyl modified poly(arylene ether sulfone) copolymer (PAES-co-APAES) was synthesized through step growth polymerization and quaternary ammonium groups were added to the backbone through post-polymerization functionalization to make PAES-co-QAPAES random copolymers. The polymer was deposited on PES membranes using a spin coating machine and ion exchanged with bicarbonate, an active ion in the MSS process. The CO_2 uptake and release performance was evaluated utilizing an inhouse made MSS setup. Molecular weight of polymers was measured using size exclusion chromatography (SEC) and the chemical structure was analyzed through nuclear magnetic resonance (1H NMR) spectroscopy. Polymer with Mn of 21.7 KDa, Mw of 43.1 KDa and dispersity (Đ) of 1.99 was synthesized, and ion exchange capacity (IEC) of 0.87 mmol/g was measured with structural analysis. Differential scanning calorimetry (DSC) confirmed random copolymer amorphous morphology with a Tg of 164 °C, thermogravimetric analysis (TGA) showed good thermal stability with T5% of 241 °C and the polymer showed relatively low water uptake of 8 wt.%. Through a humidity swing of 5-30 ppt (20%-100% RH), 114 µmol/g CO_2 uptake with a flux of 0.03 µmol/g/s and 2.96 µmol/m2/s was observed for the TFC.

10:00 am (Room Palm CF, Monday)

Oral 8 - Integrated facilitated transport membrane modules for highly selective syngas purification and carbon capture

Yang Han (The Ohio State University)*; Yutong Yang (The Ohio State University); Ruizhi Pang (The Ohio State University); Winston Ho (OSU)

Amine-containing facilitated transport membranes (FTMs) are of great interest for syngas purification. Herein, we report a new method to enhance the CO_2 permeance by manipulating the steric hindrance of the amine carrier. A series of α-aminoacids with different alkyl or hydroxyethyl substituents were deprotonated by 2-(1-piperazinyl)ethylamine, resulting in nonvolatile aminoacid salt carriers with different degrees of steric hindrance. In the presence of moisture, a bulkier alkyl substituent increased the steric hindrance and hence destabilized the carbamate adduct to afford bicarbonate through hydrolysis. This drastically increased the chemisorption of CO_2.

The enhanced CO_2 solubility significantly mitigated the carrier saturation, and an unprecedented CO_2/H_2 selectivity greater than 125 was demonstrated at 107°C and 13.8 bar of CO_2 partial pressure (Type I). As the CO_2 partial pressure reduced to 1.1 bar, a less hindered amine yielded a higher reactive diffusivity of CO_2, resulting in a CO_2 permeance of 217 GPU with a selectivity greater than 268 (Type II). These two types of FTMs were successfully scaled up by a roll-to-roll continuous coating machine. The 14″ wide scale-up membranes were then rolled into two prototype spiral-wound modules containing 800 and 1600 cm2 of the Type I and Type II membranes, respectively. At 90% CO_2 capture, the Type I and Type II modules achieved a CO_2 purity of ca. 95%, but the H_2 recoveries were only 98.2% and 97.1%, respectively. However, when the two modules were connected in series with the Type I membrane treating the syngas near the feed inlet (i.e., high CO_2 partial pressure) and the Type II membrane separating the gas in the proximity of the retentate outlet (i.e., low CO_2 partial pressure), the H_2 recovery was improved to 99.3%. The integrated membrane array also demonstrated 200-h stability under simulated syngas.

10:30 am (Room Palm CF, Monday)

Oral 9 - Thin-film Composite Membranes Based on Hyperbranched Poly(ethylene oxide) for CO_2/N_2 Separation

Gengyi Zhang (The State University of New York at Buffalo)*; Thien N Tran (University at Buffalo, The State University of New York); Liang Huang (The State University of New York at Buffalo); Erda Deng (State University of New York at Buffalo); Adrienne Blevins (Materials Science and Engineering Program, University of Colorado Boulder); Wenji Guo (University at Buffalo, The State University of New York); Yifu Ding (Membrane Science, Engineering, and Technology (MAST) Center, Department of Mechanical Engineering, University of Colorado Boulder); Haiqing Lin (The State University of New York at Buffalo)

Cross-linked amorphous poly(ethylene oxide) (XLPEO) is one of the leading membrane materials for post-combustion CO_2 capture. For example, XLPEO prepared from poly(ethylene glycol) methyl ether acrylate (PEGMEA) exhibited CO_2 permeability of 570 Barrer and CO_2/N_2 selectivity of 41 at 35℃. However, these XLPEOs cannot be dissolved in coating solutions, making it impossible to be fabricated into thin-film composite (TFC) membranes using state-of-the-art manufacturing processes. In this study, we synthesized high molecular weight yet soluble HPEO via atom transfer radical polymerization (ATRP). HPEO with molecular weights from 180 kDa to 220 kDa were synthesized by adjusting the ratio of the PEGMEA and initiator (pentaerythntol tetrakis(2-bromo-isobutyrate)). All polymers are amorphous at 35 ℃ and show a similar glass transition temperature of around -66oC. Increasing the polymer molecular weight from 180 to 220 kg/mol decreases the CO_2 permeability from 540 to 450 Barrer while retaining CO_2/N_2 selectivity of 43. Gas solubility is independent of the polymer molecular weight. A polymer with the best combination of CO_2 permeability (540 Barrer) and CO_2/N_2 selectivity (43) was fabricated into TFC membranes by a solution casting method with a gutter layer of polydimethylsiloxane (PDMS). The gutter layer shows CO_2 permeance of 9000 GPU, while the PEO-based TFC membranes exhibit CO_2 permeance of 1000 GPU with CO_2/N_2 selectivity of 38 at 35°C. When challenged with simulated flue gas containing water vapor at 35℃ for over 100 h, the membrane shows stable CO_2 permeance of 850 GPU and CO_2/N_2 selectivity of 37, demonstrating their potential for carbon capture. We will also present the effect of small molecular additives in the HPEO (such as crown ethers and polyethylene glycol dimethyl ether) on the CO_2/N_2 separation properties.

11:00 am (Room Palm CF, Monday)

Oral 10 - Polyvinylpyridine-based Catalytic Membranes for Integrated CO_2 Capture and Conversion: Structure-function Relationships

Casey O'Brien (University of Notre Dame)*

Current sorbent-based CO_2 capture processes are prohibitively expensive because they require energy intensive regeneration stages to release CO_2 and regenerate the sorbent. Additionally, they produce CO_2 streams that must be compressed and transported to sites for storage, which reduces efficiency. We are developing a catalytic membrane technology that addresses these challenges and potentially enables commercially viable CO_2 capture by integrating CO_2 capture and CO_2 conversion to value-added chemicals in a single unit process operated continuously at mild conditions. Such a membrane would act as both the CO_2 separation and conversion medium, providing an energy- and atom-efficient alternative to sorbent-based CO_2 capture, compression, transport, and storage. We have recently discovered that polyvinylpyridine (PVP)-based membranes are promising for demonstrating the potential of amine-functionalized membranes as an integrated CO_2 capture and conversion platform tunable for CO_2 conversion to a variety of products. In this talk, I will demonstrate that PVP-based membranes are permeable and selective CO_2 separation membranes that are also catalytically active for cyclic carbonate synthesis at mild conditions (23 °C, 1 atm CO_2). The performance of PVP for CO_2 separation and for catalytic conversion of CO_2 can also be enhanced by quaternization of PVP with alkyl halides or by metalation with various metal salts or metal nanoparticles. Finally, I will demonstrate that PVP-based membranes can integrate CO_2 capture from a simulated flue gas and convert it with epoxides to cyclic carbonates in a separate stream. Integrating CO_2 capture and conversion could be a transformative shift in carbon capture technology that could also revolutionize the way chemicals are produced.

11:30 am (Room Palm CF, Monday)

Oral 11 - Effect of free volume modification (FVM) on the physical aging behavior of amine-functionalized PIM-1 membranes for gas separation

Taigyu Joo (Massachusetts Institute of Technology)*; Katherine Mizrahi Rodriguez (MIT); Hyunhee Lee (MIT); Zachary Smith (MIT)

Multiple post-synthetic functionalizations of PIM-1 have been studied to enhance CO_2 sorption for gas separation applications. While functionalizing PIM-1 with polar moieties increases CO_2 sorption, introducing secondary interactions results in a concomitant decrease in free volume, reducing the overall membrane performance. One way to overcome this issue is by using a free volume modification (FVM) method. The FVM method utilizes labile functional groups, such as tert-butoxycarbonyl group, to alter packing structures and transport properties of polymers. By thermally removing the functional groups in the solid-state, new free volume elements are introduced and a narrower free volume distribution is produced. This method was demonstrated to be extremely effective for amine-functionalized PIM-1 (PIM-NH_2), in which new free volume elements are added along with light urea crosslinks to simultaneously boost permeability and selectivity. However, higher free volume makes membranes more susceptible to physical aging. In this study, the physical aging of modified PIM-NH_2 (PIM-deBOC) is closely examined using permeation and sorption tests of non-plasticizing gases over approximately 1000 h. Despite having the urea crosslinks for stabilization, PIM-deBOC showed faster reductions in permeability compared to the control PIM-NH_2, and the reduction primarily arose from decreases in diffusivity. Correspondingly, PIM-deBOC also showed the largest increases in selectivity. The examination of packing structures suggested that the crosslinks in PIM-deBOC increase selectivity, but are insufficient to stabilize high free volume samples from physical aging effects. This study provides a comparison of aging behavior for over 1000 h for various compositions of the PIM subclass, including nitrile, amine-functional, and urea crosslinked samples and provides guiding principles on the role of free volume, hydrogen bonding, and crosslinking on physical aging behavior for microporous polymers.

12:00 pm (Room Palm CF, Monday)

Oral 12 - The Effect of Channel Height Variation on CO_2 Separation Performance in a Flat Sheet Membrane

Joanna Rivero (University of Pittsburgh)*; Marcos Da Conceicao (University of Toledo); Leo Nemetz (University of Toledo); Glenn Lipscomb (University of Toledo, Toledo); Katherine Hornbostel (University of Pittsburgh)

Membranes are commonly used for CO_2 separation. Flat sheet membranes offer advantages over other membrane configurations, such as ease of manufacturing and low pressure drop. Our collaborators at Membrane Technology & Research (MTR) have moved towards a flat sheet design in their latest Gen 2 Polaris membrane module, which is currently undergoing pilot-testing at Technology Centre Mongstad. One potential disadvantage of using flat sheet membranes is that the manufacturing process can lead to variability in gas channel height, leading to a decrease in overall performance and gas separation rates. In this work, the impact of channel height distribution on gas separation performance of flat sheet carbon capture membranes is investigated. A 1D steady-state model of a flat sheet membrane was developed in MATLAB® that simulates the mass transfer and pressure drop along the length of each flow channel. Gas channel heights are modeled using a Gaussian distribution, then the flow-rate distribution is calculated for fixed pressure conditions. The model has already been verified against a published ideal hollow fiber membrane model at high recovery rates, where membrane configuration has a negligible effect on results. We are also planning to validate our model against in-house experiments at the University of Toledo with synthetic flue gas and a small flat sheet membrane. With the channel height variation study, a drop in overall performance is noted as channel height variation increases. In addition, we compare our flat sheet membrane model against a hollow fiber membrane model that incorporates a similar fiber size distribution. System-level modeling will also be performed in Aspen Custom Modeler to predict how channel size variation impacts overall system performance for MTR's pilot-scale test set-up. The model developed here will build on previous ideal flat sheet models and enable better prediction of performance by accounting for non-ideal effects.

Oral 3 – EMERGING MATERIALS FOR LIQUID SEPARATIONS I

9:30 am (Room Palm BE, Monday)

Oral 13 - Separation of Trace Organics from Aqueous Streams using Carbonaceous Membranes

Haley D White (Georgia Institute of Technology)*; Ryan Lively (Georgia Tech)

The need for energy-efficient aqueous waste stream reclamation technologies is becoming more urgent as global population, energy demand, and water consumption continue to increase. Many of these aqueous waste streams contain recalcitrant organic contaminants, such as pharmaceuticals, industrially used solvents, and personal care products, that must be removed prior to reuse. We hypothesize that rigid carbonaceous materials can better remove organic contaminants due to an increase in entropic selectivity relative to separations involving flexible membrane materials. Here, fundamentals in transition state theory and thermodynamics are used to develop a model that can predict the separations of example organic-laden aqueous waste streams in reverse osmosis processes using rigid nanoporous materials. The contributions of entropic selectivity and isotherm behavior are estimated and correlated with predicted separation performances. Then, bench-scale reverse osmosis and pervaporation experiments are conducted with a solution of n, n- dimethylformamide in deionized water to assess the separation performance of carbonized membranes derived from a polymer of intrinsic microporosity 1 precursor. Structural and elemental characteristics of carbonized membranes are probed with nitrogen physisorption, x-ray photoelectron spectroscopy, and scanning electron microscopy; these results provide a deeper understanding of transport phenomena in separations using rigid materials. Benefits and limitations of using rigid membrane materials in aqueous separations are discussed, and a recommended path forward for using such materials for aqueous waste stream reclamation is provided.

10:00 am (Room Palm BE, Monday)

Oral 14 - Cucurbit[n]uril-based nanofiltration membranes for effective molecular separation

Shi-Peng Sun (Nanjing Tech University)*

Thin-film composite (TFC) nanofiltration (NF) membranes have been widely employed in seawater desalination, heavy metal ion removal, and other water purification processes. Unfortunately, the inherent ""trade-off"" effect between permeability and selectivity hampers their progresses in the fine separation of small molecules with similar molecular weights. Herein, a rarely reported cucurbit[n] urilis employed to construct the TFC NF membrane, which has profound effects on both membrane structures and separation performance. On one hand, CBn can inhibit the diffusion of aqueous monomers in the interfacial polymerization by its large intrinsic cavity, inducing the controllable and quantifiable nanostrand structures on the membrane surface, which is important for permeability improvement. On the other hand, the CBn ports have characteristic separation sizes and rich carbonyl groups to simultaneously enhance the size sieving and Donnan effects of NF membranes, which is helpful for the separation of small molecular mixtures with similar molecular weights. Therefore, it has great potential in breaking the ""trade-off"" effect of selectivity and permeability. We have found CBn is promising in tuning the membranes to dramatically enhance the water permeability while maintain or improve the selectivity in sulphate/chloride, isopropanol/water separations. The membranes also exhibit antifouling and self-cleaning properties, which are promising in water treatment, desalination and other applications.

10:30 am (Room Palm BE, Monday)

Oral 15 - Nanoporous multilayer graphene membrane for precise and stable forward osmosis desalination

Claudio A Ruiz Torres (The University Of Toronto)*; Jay Werber (University Of Toronto).

Graphene oxide (GO) membranes have shown exceptional performance in fluid separation processes due to their intrinsic properties and abundant charges on oxygen-containing groups. However, in desalination, swelling of GO laminates in water promotes the enlargement of interlayer spacing (d) to ~13.5 Å, restricting the ability to retain common salts. Several approaches have been suggested to suppress the interlayer spacing expansion, including physical confinement, crosslinking, or partial reduction, but stable and high salt rejection rate in long-term operation time is rarely realized in a scalable and practical manner. Swelling restriction by controlled oxidation of the laminates enhances the retention of mono and divalent ions. Solvent permeation, on the other hand, also highly depends on the interlayer spacing. Thus, the presence of in-plane nanopores in graphene-based membranes enhances solvent permeation, which is highly essential for an efficient desalination process. In this work, oxidation-controlled nanoporous graphene (OCNG), was prepared by graphite intercalation chemistry. Multilayer OCNG membranes were analyzed and tested in long-term forward osmosis desalination. The presence of nanopores was confirmed by electron microscopy. OCNG is dispersible in water; thereby, membrane fabrication is easily achieved on porous supports by vacuum filtration. The controlled oxidation in the laminates shows a drastic reduction in channel swelling (9.12), reducing ion permeation 5-fold in comparison with fully oxidized graphene oxide. Using FO mode, our membrane demonstrated great performance in the removal of the most common ions in seawater (>99%), including in harsh conditions, hypersaline feeds, and 7 continuous days of operation. In long-term forward osmosis desalination (month-scale), OCNG membrane notably surpassed most of the novel water desalination membranes, showing a clear advantage in terms of stability and efficient high-salinity water purification.

11:00 am (Room Palm BE, Monday)

Oral 16 - High flux vapor-gap membranes for reverse osmosis via informed design of membrane pore structures

Elizabeth A Hjelvik (University of Colorado Boulder)*; Kian P Lopez (University of Colorado Boulder); Anthony Straub (University of Colorado Boulder)

Current RO membranes are limited in their ability to remove low molecular weight contaminants and degrade when exposed to strong oxidants commonly used in water treatment. Vapor-gap RO is an emerging water desalination technology with the potential to address the current drawbacks of RO. We have previously shown that this technology allows for complete removal of nonvolatile contaminants that are poorly rejected by conventional RO membranes and, the hydrophobic membranes used for this process resist degradation from strong oxidants. Despite having improved selectivity and oxidation resistance, the permeability of water through vapor-gap processes is orders of magnitude lower than typical RO system. Given that transport in the membrane is dominated by Knudsen diffusion, which is reliant upon pore radius and membrane thickness, we hypothesize the low water permeability can be addressed through informed design of membrane pore structure and size. In this work, we explore the application of asymmetric pore membranes as a solution to vapor gap RO's low water permeability. The inclusion of two radii in the pore structure allows for two key benefits: (1) a <2 µm thick surface layer with a <10 nm pore size enables high pressure operation (up to 120 bar) without pore wetting and (2) a thicker underlying large pore radius layer minimizes Knudsen transport resistances and allows for 20 times higher water fluxes in comparison to symmetric pores. To inform the optimal pore design, transport models have been constructed to predict the flux through the membranes. This model is validated experimentally through the modification of ceramic membranes with tunable pore structures via chemical vapor deposition. This work serves as a transformative step for understanding how materials properties can help design high flux and high selectivity membranes for future applications.

11:30 am (Room Palm BE, Monday)

Oral 17 - Preparation of functionalized porous poly (ether ether ketone), PEEK, materials and hollow fiber membranes

Andrii Bazylevich (Ariel University); Sagiv Weintraub (Ariel University); Gary Gellerman (Ariel University); Ben Bikson (Avanpore)*

Commercial application of organic solvent nanofiltration, OSNF, is currently limited due to limited solvent and thermal resistance of commercial polymeric membrane materials. To rival the industrial success of membranes in water treatment applications advances in polymeric membrane chemical and solvent stability are needed. Poly(ether ether ketone) (PEEK) is commercial engineering polymer with almost universal solvent and chemical resistance and outstanding thermo-mechanical properties. These properties make PEEK ideal material for preparation of OSNF membranes. A novel method of forming porous membranes from PEEK using a modified TIPS process was recently developed and used to prepare hollow fiber membranes [1]. Of particular interest are PEEK hollow fiber membranes with pore sizes in the nanometer range. Surface functionalization of preformed porous PEEK articles can be used to impart additional tailored separation properties. Since PEEK polymer is intrinsically hydrophobic, development of surface functionalized PEEK membranes with hydrophilic characteristics is of interest. Functionalization of preformed porous PEEK materials must be carried out without affecting pore morphology. Traditional methods of PEEK article surface functionalization are aggressive and affect crystalline phase leading to loss of chemical/solvent resistance We report a novel method of functionalizing preformed mesoporous PEEK materials with hydroxyl groups. The pore structure of functionalized materials was characterized as a function of the degree of substitution. The methodology was applied to preparation of asymmetric PEEK hollow fiber membranes wherein alternatively bulk pore surfaces or the surface layer only were functionalized with hydroxyl groups. A new method of characterization of the degree of PEEK surface modification by -OH groups was developed.

References

1. Y. Ding and B. Bikson, J. Membr. Sci., 357 (2010) 192

12:00 pm (Room Palm BE, Monday)

Oral 18 - Polyethyleneimine-based Surface Treatment for Facile Synthesis of Conductive Membranes: Active Material Doesn't Matter

Wei Zhang (University of North Carolina at Chapel Hill)*; Orlando Coronell (University of North Carolina at Chapel Hill)

Electrically conductive membranes have been demonstrated in the literature as a promising tool to enhance the performance of membrane-based water/wastewater treatment technologies. Membrane surface functionalization with "active" conductive materials is a direct and effective approach to obtain membranes with electrically conductive properties. However, a surface functionalization strategy that can be used for most types of commercial membranes (e.g., reverse osmosis, nanofiltration, ultrafiltration, and microfiltration) is not available yet. To address this need, we report a facile and low-cost polyethyleneimine-based crosslinking method that enables rendering electrically conductive the surfaces of a broad range of commercial membranes (i.e., SWC 4+, ESPA 3, NF 270, PSf 20 KDa, and 0.1 μm PVDF membranes) by using graphite. All synthesized graphite composite membranes showed low sheet resistance (< 3 kΩ/sq), improved separation performance (i.e., solute rejection) and reduced water permeance compared with the insulating pristine membranes. The electrical conductivity of the graphite-coated conductive membranes was demonstrated through 100% degradation of methylene blue (30 mg/L) in 50 min with a 5 V applied potential. It was demonstrated that the PEI-based membrane crosslinking method is suitable for "active" conductive materials other than graphite, including carbon nanotubes, reduced graphene oxide, activated charcoal, and silver nanoparticles. Taken together, this work presents a promising strategy for the synthesis of electrically conductive membranes with a variety of "active" materials for various environmental applications.

Oral 4 – MEMBRANE SYNTHESIS AND CASTING I

9:30 am (Room Abbey North, Monday)

Oral 19 - Microstructure evolution during nonsolvent-induced phase separation: a joint experimental-computational investigation into membrane formation

Alexander Bridge (The University of Texas at Austin)*; Anthony J Cooper (University of California, Santa Barbara); Jan Garcia (UC Santa Barbara); Matthew Santoso (The University of Texas at Austin); Glenn Fredrickson (UC Santa Barbara); Benny Freeman (The University of Texas at Austin)

Membrane-based separation processes favor asymmetric structures prepared via nonsolvent-induced phase separation (NIPS). Although a wealth of experimental knowledge has facilitated the development of application-optimized membrane structures, our mechanistic understanding of NIPS remains largely qualitative. This study reports a joint experimental-computational investigation into NIPS membrane formation using model systems based on polysulfone (PSf). Asymmetric membrane structures (i.e., a thin "active" skin layer surmounting a porous sublayer), characterized via electron microscopy and image segmentation, were analyzed in relation to important experimentally-measured parameters (e.g., binary thermodynamic interactions between system components and concentration-dependent polymer-solvent viscosities) for several ternary PSf-solvent-coagulant membrane-forming systems. To expand upon experimental insights, a phase field model was used to simulate phase inversion of the PSf-solvent-coagulant systems using experimentally-quantified parameters. The effects of changes in specific parameters on phase behavior were simulated and analyzed in relation to imaged membrane structures. Simulated membrane structures accurately reflect experimental observations and provide unique insights, previously inaccessible via experiment, into skin layer/sublayer formation and transitions from finger-like to spherical sublayer pores.

10:00 am (Room Abbey North, Monday)

Oral 20 - Automated stepwise synthesis of poly(acrylonitrile)-supported oligoamide membranes and their dye–salt separation

Paramita Manna (Ben-Gurion University of the Negev); Roy Bernstein (Ben Gurion University); Ron Kasher (Ben Gurion University of the Negev)*

Ultrafiltration (UF) and nanofiltration (NF) membranes are widely used in separations in industrial processes such as potable water purification, the food and the pharmaceutical industries, the dye industry, and hemodialysis. However, producing specialized membranes with diverse transport properties is challenging. A method for fabricating membranes with controllable transport properties and its automation are described here by stepwise synthesis (molecular-layer by layer) of aromatic oligoamide using trimesoyl chloride and meta-phenylenediamine monomers on poly(acrylonitrile) (PAN) supports. Optimization of the synthesis parameters yielded tight UF membranes with excellent transport properties: A membrane with 10.5 oligoamide synthetic cycles (OA-10.5) had water permeance of 88 L·m−2h−1 bar-1, and molecular weight cutoff (MWCO) of 6.65 kDa. An automated synthesis method for the fabrication of the oligoamide membranes was developed using an automated peptide synthesizer. The oligoamide membranes showed good retention of water-soluble organic dyes, and could separate organic dyes from salt: The membrane with 10.5 synthetic cycles, OA-10.5 membrane, showed 100% rejection of Congo red in the presence of 20 g/L NaCl, with >98% NaCl passage [1]. The OA-10.5 membrane kept this separation performance in long-term filtration of 100 h, which proves the excellent long-term stability of these oligoamide membranes. Thus, the prepared oligoamide membranes hold great promise for dye–salt separations that are highly important for increasing the product recoveries in the dye industry.

References

[1] Manna, P.; Bernstein, R.; Kasher, R., Stepwise synthesis of polyacrylonitrile-supported oligoamide membranes with selective dye–salt separation. J. Membr. Sci. 2022, 643, 120035.

10:30 am (Room Abbey North, Monday)

Oral 21 - Microporous polymer membranes: a new approach to simultaneously reducing physical aging and swelling while enhancing permeability

Jing Deng (University of Oklahoma); Cara Doherty (CSIRO); Laura Matesanz Nino (University of Valladolid); Angel Lozano (University of Valladolid); Cristina Alvarez (University of Valladolid); Anita J. Hill (CSIRO); Michele Galizia (University of Oklahoma)*

Although exhibiting superb performance in the short time operation, microporous and ultra-microporous glassy polymers suffer from physical aging and swelling, which hampers their scale up to the industrial level in gas, vapor and organic solvent separations. In this study, we propose a facile approach to manipulate free volume morphology and distribution, as well as polymer chain dynamics, to reduce physical aging and swelling in microporous glassy polymers, while enhancing gas permeability. A model microporous high free volume glassy polymer, PTMSP, was blended with 5wt.% hyper cross-linked triptycene-isatin porous polymer network exhibiting configurational free volume. As a result, gas permeability increased by a factor of three compared to neat PTMSP, with a minor loss in ideal selectivity. The combination of PALS analysis, density measurements and free volume calculations indicate that blending with PPN provides a unique opportunity to exquisitely tune the size and architecture of free volume elements, leading to the above mentioned permeability enhancement, as well as to a remarkable reduction of aging rate and swelling propensity. The molecular mechanism by which these effects take place was elucidated in detail, and possible strategies to enhance selectivity other than permeability and long-term stability were devised.

Remarkably, although the incorporation of PAFs (Porous Aromatic Frameworks) in high free volume glassy polymers can successfully slow down the aging rate while enhancing gas permeability and retaining selectivity, their cost is very high and their synthesis pathway complicate, which hampers their industrial scale-up. In sharp contrast, the PPNs synthesis pathway is easy and one order of magnitude cheaper, which makes our approach highly scalable.

11:00 am (Room Abbey North, Monday)

Oral 22 - Crosslinked Triptycene-containing Polybenzoxazole Membranes with Model Network Structures for High Performance Gas Separations

Ziwei Dai (University of Notre Dame)*

Crosslinked polymeric membranes have been shown to address problems associated with polymeric membranes such as physical aging and condensable-gas-induced plasticization for gas separation applications. However,common random crosslinking approaches generally have minimal control and limited tunability in microstructure, leading to unpredictable transport properties and significantly reduced permeabilities.This study provides a novel cross-linking technique that permits fine control of microstructures and customizable membrane characteristics using telechelic oligomers with well-controlled molecular weight to generate model network polymer membranes.Specifically, a series of bimodal crosslinked triptycene-containing poly(benzoxazole)-based membrane was fabricated with a short and a long oligomer chain to achieve systematically controlled crosslink inhomogeneity and density.Upon heat treatment, thermal-induced crosslinking reaction was only allowed to happen at chain ends of oligomers and by manipulating molar fractions of respective components, overall inter-crosslink molecular weights in fabricated membranes were controlled. The use of a hierarchical triptycene-based PBO backbone (TPBO) in combination with bulky phenyl groups at crosslink sites counteracts the undesirable densification caused by crosslinking, resulting in highly permeable and selective crosslinked membranes that outperform most existing crosslinked systems. All bimodal membranes displayed general improvements in gas permeabilities compared with the unimodal counterparts. These crosslinked bimodal TPBO films approached or surpassed the 2008 Robeson upper bound for H_2/CH_4, CO_2/CH_4, and O_2/N_2 separation, suggesting their promising potentials for applications in H_2 and CH_4 purification and air separation.The synthesis and characterization of these crosslinked polymer membranes will be discussed in this talk, with an emphasis of underlying connections between microstructures and gas transport properties.

11:30 am (Room Abbey North, Monday)

Oral 23 - Integrated Roll-to-Roll (R2R) Fabrication of Graphene Oxide Nanofiltration Membranes for Applications in Extreme Operating Environments

Amirsalar Rabbani Esfahani (Georgia Institute of Technology)*; Chen Ma (Georgia Institute of Technology); Uwezo Flewellen (Georgia Institute of Technology); Sankar Nair (Georgia Institute of Technology); Tequila Harris (Georgia Institute of Technology)

Economical fractionation of multicomponent biomass-derived streams is a key challenge. Kraft black liquor (BL) is generated at ~1 billion tons/yr globally from biomass pulping, and contains ~15 wt% total solids including lignin, hemicellulose fragments, and inorganics. Membrane-based BL concentration [1] is attractive, but conventional membranes are precluded by low solute rejections and poor stability in BL, which combines alkaline pH (~13) and high temperature (70-85°C). Georgia Tech has pioneered the development and pilot trials of robust graphene oxide (GO) membranes for kraft BL processing [2-4]. Here we discuss the scaled fabrication and GO nanofiltration membranes by a dynamic/roll-to-roll (R2R) process integrated with vacuum filtration. This process uses only aqueous suspensions, and no organic chemicals. We have fabricated reduced GO (rGO) membranes [3] and rGO-X membranes, in which X = a large □-conjugated polyaromatic species that binds strongly to the rGO sheets and dramatically changes its characteristics [4]. We discuss the key dynamic processing factors that determine the quality of the R2R membranes and then optimize the R2R process on sheet sizes up to 3 ft (L) x 1 ft (W). We obtain high-quality, "defect-free" GO membranes with thicknesses in the range of 100-200 nm, and present additional characterizations of their microstructure and uniformity. We piloted these R2R membrane sheets for extended periods (weeks) under cross-flow conditions with real BL feeds. We present detailed separation results indicating that R2R-fabricated membranes have equivalent performance as their static-fabricated counterparts. These results have positive implications for the industrial applications of GO membranes.

References: [1] Kevlich et al, Sep Sci & Tech (2017). [2] Rashidi et al, ACS Sus Chem Eng (2017). [3] Wang et al, ACS Sus Chem Eng (2019). [4] Wang et al, Nature Sus (2021).

12:00 pm (Room Abbey North, Monday)

Oral 24 - Super-hydrophobic MFI Zeolite Hollow Fiber Membranes for Alcohol/Water Separation

Seyed Mirfendereski (Arizona State University); Jerry Lin (Arizona State University)*

Most zeolite membranes are hydrophilic, and thus water perm-selective. It is highly desirable to prepare alcohol selective membranes for many applications including bioseparation. This paper reports a zeolite membrane synthesis method that increases the hydrophobicity of zeolite layer and thus alcohol selectivity of the membrane by enhancing the Si/Al ratio and crystallinity without scarifying the permeability through decreasing the membrane thickness. The method uses a solid template for preparing a hierarchical mesoporous silicalite zeolite layer as an "intermediate mesoporous zeolite layer (IMZ)" between alumina support and zeolite layer. This mesoporous layer acts as a barrier that effectively terminates the infiltration of Al ions from the alumina support into the zeolite reaction mixture during the synthesis of top layer. It also provides a permeable and mechanical resisting support for the thin zeolite top layer, due to preparing a higher nucleation rate than the crystal growth rate. The varying-temperature secondary growth then completes the formation of the composite zeolite layer by an ultra-thin nonporous layer. The effects of primary seeding method, seed particle size, seed size ratio, and variable temperature/time profile on the microstructure and hydrophobicity of the hierarchical mesoporous layer are investigated leading to identification of optimum conditions for synthesis of thin and uniform intermediate layer and final zeolite layer with high alcohol/water separation performance. The ultimate high-performance MFI zeolite membranes have microstructure consisting of an alumina support, a thin mesoporous intermediate layer, and an ultra-thin, dense, and fully inter-grown top zeolite layer. The best hollow fiber supported MFI zeolite membrane with a Si/Al ratio of 187 exhibits methanol/water or ethanol/water pervaporation separation factor of 160 with total flux of 5 (for methanol/water) and 2.9 (ethanol/water) kg m^{-2} h^{-1}, respectively.

Oral 5 – INORGANIC MATERIALS

9:30 am (Room Abbey South, Monday)

Oral 25 - High-pressure CO_2 permeation properties of ceramic-carbonate dual-phase membranes

Oscar Ovalle (Arizona State University); Jerry Lin (Arizona State University)*

This paper reports carbon dioxide (CO_2) permeation properties and stability of ceramic-carbonate dual-phase membranes at high pressures/temperatures which are critical to their CO_2 separation and membrane reactor applications. Experiments were conducted to study the effect of high transmembrane pressure on CO_2 permeation flux and the stability of molten carbonate in the samarium-doped ceria (SDC) and molten-carbonate membranes. Porous SDC tubular supports were made by a cold isostatic press (CIP)/sintering method with low porosity (below 7%), and gas-tight SDC-MC membranes were prepared by direct infiltration of molten lithium and sodium carbonate mixture into SDC pores, with a molten-carbonate volume fraction less than 7%. CO_2 permeation/ separation tests were performed on the SDC-MC membranes using feed gas of equal molar CO_2/N_2 mixture at feed pressures up to 15 atm and sweep gas of helium at 1 atm. CO_2 permeation fluxes for these membranes can be described by a mathematical model which shows logarithmical feed/permeate CO_2 pressure ratio dependence of CO_2 permeation flux in 660–810 °C with feed CO_2 partial pressures up to 7.5 atm (total pressures up to 15 atm). The temperature dependence of CO_2 permeation flux data gives activation energy of 30 kJ/mol. The SDC-MC membranes exhibit a low effective carbonate conductivity due to low molten-carbonate volume fraction, leading to CO_2 permeation dominated by carbonate ionic conduction in the MC phase. MC in the SDC-MC membranes stays in the pores of SDC support of the membranes after high-temperature permeation/separation tests with transmembrane pressure up to 15 atm because of the high capillary pressure holding the molten carbonate. SDC-MC membranes show high stability for high-temperature, high-pressure separation, and chemical reaction applications.

10:00 am (Room Abbey South, Monday)

Oral 26 - Metal ions-doped ultrahigh flux composite graphene oxide/carbon nanotubes (GO-SWCNT) membranes for efficient H_2 separation

Daniel Gardeno (University of Chemistry and Technology Prague); Zdeněk Sofer (University of Chemistry and Technology Prague); Daniel Bouša (University of Chemistry and Technology Prague); Saeed Jamali Ashtiani (University of Chemistry and Technology Prague); Karel Friess (University of Chemistry and Technology Prague)*

We report on the preparation, modification and H_2 separation performance of ultrahigh flux composite GO-SWCNT membranes. Thin selective GO layers[1] were deposited on the top of ultrapermeable SWCNT supports [2]. The resulting combination led to higher H_2/CO_2 selectivity compared to previously reported self-standing GO membranes [1] but with substantially higher fluxes and dramatically better mechanical and time-stability. Fabricated GO-SWCNT membranes exhibited remarkable gas separation performance towards H_2/CO_2, H_2/CH_4, H_2/N_2 and N_2/CH_4 gas pairs that far exceed the corresponding 2008 Robeson upper bound [3], especially after the doping by selected metal ions. Such doping caused specific structure/properties modification of GO-SWCNT and improved substantially gas permeability and selectivity. Presented composite GO-SWCNT membranes demonstrate a promising route towards large-scale fabrication of high flux hydrogen-selective gas membranes intended for H_2/CO_2, H_2/CH_4 and H_2/higher alkanes separations.

Acknowledgements

The Czech Ministry of Education financed this research, Youth and Sports, by project LTAUSA19038 and specific university research (MŠMT No. 20-SVV/2021 and 2022)

References

[1] D. Bouša et al., Chemistry-A European Journal, 23 (2017) 11416-.

[2] M. Boháčová et al., Applied Materials Today 15 (2019) 335-342.

[3] L. M. Robeson, Journal of Membrane Science 320 (2008) 390-400.

10:30 am (Room Abbey South, Monday)

Oral 27 - Network nanostructured ZIF-8 to enable percolation and enhanced gas transport

Hyunhee Lee (MIT)*; Won Seok Chi (Chonnam National University); Moonjoo Lee (Samsung); Ke Zhang (Aramco Americas); Katherine Mizrahi Rodriguez (MIT); Stephen DeWitt (MIT); Zachary Smith (MIT)

Inorganic materials such as MOFs are important for high-performance gas separation membranes with good stability. Unfortunately, MOF crystals are difficult and expensive to process into modules. Instead, MOFs are often incorporated into polymers to form MMMs, where a percolation network between MOF fillers could enable high-performance. However, particles tend to aggregate, preventing the MOF network in the polymer matrix from forming and even resulting in non-selective defects. Thus, even at high MOF loadings, particle percolation, and hence, MOF-like transport is rarely accessible.

Here, a facile method to synthesize branched ZIF-8 (BZ) nanoparticles at ambient conditions has been developed for efficient gas separations that can achieve this percolation network. Due to the small particle dimensions of BZ and the high surface area–to–volume ratios, ZIF-8 ligand motion is suppressed when combining into the polymer–ZIF composites. This gate opening suppression results in exceptionally high selectivities for small gases (e.g., H_2-based) compared to the typical propylene/propane size cut-off observed for traditional ZIF-8 control (CZ) particles. Additionally, mechanical property analysis and NMR relaxation studies indicate that the BZ morphology effectively stabilizes polymer chains, preventing plasticization up to 50 bar. Most significantly, the novel BZ morphology readily forms a MOF percolation network within the polymer matrix, resulting in MMMs with up to 85% higher hydrogen flux than those formed with traditional CZ materials. The triple benefit of a percolation network, confinement of MOF ligand gate opening, and enhanced polymer-phase stability present a unique opportunity to address multiple challenges in applying MOFs in mixed-matrix membrane (MMM) form factors, resulting in separation performance that surpasses Robeson's upper bound for H_2-based separations and exceptional stability under mixed-gas conditions.

11:00 am (Room Abbey South, Monday)

Oral 28 - High-performance CO_2- and O_2-sieving by controlled etching of graphene

Shiqi Huang (Ecole polytechnique fédérale de Lausanne)*; SHAOXIAN LI (EPFL); Luis Francisco Villalobos (Yale University); Kumar Varoon Agrawal (Ecole polytechnique fédérale de Lausanne)

Nanoporous graphene is a highly promising selective layer for energy-efficient gas separation. With an extremely low diffusion resistance across the atom-thick nanopores, the N-SLG membrane can yield large CO_2 and O_2 permeance and an attractive gas pair selectivities [1-3]. However, it has remained challenging to incorporate CO_2- and O_2-sieving nanopores in graphene with a narrow-enough pore-size distribution (PSD) to obtain the needed sub-angstrom resolution in differentiation from N_2, for application in carbon capture.

Herein, we developed a millisecond ozone-based carbon gasification chemistry to control nanopores incorporation in graphene lattice [4]. The rapid etching kinetics with short and controllable pore expansion time led to high-density (10E12 cm-2) nanopores with narrow PSD in N-SLG. The obtained N-SLG enabled to effectively sieve various similarly-sized gas pairs, including several industrially relevant mixtures as CO_2/O_2, CO_2/N_2 and CO_2/CH_4. Furthermore, we show the molecular cutoff can be tuned by 0.1 Å via ozone-based pore-edge functionalization and oxygen-based slow etching. The resulting N-SLG membrane reached O_2/N_2 selectivity of 3.5 with corresponding O_2 permeance of 1300 gas permeation units (GPU), and CO_2/N_2 selectivity of 21.7 with corresponding CO_2 permeance of 11850 GPU. The excellent gas separation performance for the post-combustion carbon capture will likely tilt the capture technology toward the membrane-based process.

References

(1) Wang, L. et al., Nature Nanotechnology 2017, 12, 509–522.

(2) Huang, S. et al., Nature Communications 2018, 9, 1–11.

(3) Zhao, J. et al., Science Advances 2019, 5, eaav1851.

(4) Huang, S.et al., Science Advances, 2021, 7(9): eabf0116.

11:30 am (Room Abbey South, Monday)

Oral 29 - Tunable organosilica membranes for high-temperature H_2/CO_2 separation synthesized by rapid oxygen plasma treatment of polysiloxane

Vinh Bui (University at Buffalo)*; Lingxiang Zhu (U.S. Department of Energy National Energy and Technology Laboratory); LEIQING HU (University at Buffalo); Liang Huang (The State University of New York at Buffalo); Elizabeth Haddad (University at Buffalo); Erda Deng (State University of New York at Buffalo); Thien N Tran (University at Buffalo, The State University of New York); Haiqing Lin (The State University of New York at Buffalo)

Nanoporous silica membranes with strong size sieving ability are attractive for sustainable H_2 production and precombustion CO_2 capture. However, conventional process to fabricate silica membrane is complicated and requires high-temperature calcination steps, which hinder large-scale production for industrial applications. This study focuses on a facile way to fabricate silica membranes using rapid oxygen plasma treatment of polysiloxanes at room temperature. Specifically, thin-film composite (TFC) membranes of polydimethylsiloxane (PDMS) are exposed to oxygen plasma at pressures up to 0.5 torr for 0.5 – 3 min, forming a thin organosilica layer (~10 nm) on top of the PDMS, as confirmed using SEM, AFM, and XPS. The effects of the treatment conditions on the H_2/CO_2 separation properties are thoroughly characterized, such as exposure time, oxygen pressure and flow rate, inductively coupled plasma power (ICP), and reactive ion etching power (RIE). Increasing the plasma exposure time and ICP power decreases gas permeance and increases H_2/CO_2 selectivity. On the other hand, increasing oxygen pressure increases gas permeance and decreases gas selectivity, while oxygen flow rate exhibits neglectable effect on the separation performance. Surprisingly, gas selectivity reaches a maximum at RIE power of 20 W while gas permeance decreases with increasing RIE power. The membrane prepared at RIE power of 20 W and ICP power of 200 W exhibits the best separation properties with H_2 permeance of 880 GPU and H_2/CO_2 of 67 at 150oC, far surpassing Robeson's upper bound. The effect of the polysiloxane chemistry on the separation properties of the formed membranes will also be discussed. The robust and tunable H_2/CO_2 separation properties coupled with the facile fabrication demonstrate the potential of the membrane for practical H_2 purification and CO_2 capture.

12:00 pm (Room Abbey South, Monday)

Oral 30 - Synthesis of ZIF membranes by a vapor-phase seeding method for olefin/paraffin separation

Zhiqin Qiang (University of Wisconsin-Milwaukee)*; Weiling Xia (University of Wisconsin-Milwaukee); Rachel Clark (University of Wisconsin-Milwaukee); Zihao Yi (University of Wisconsin-Milwaukee); Rahul Sampat Khandge (University of Wisconsin-Milwaukee); Xiaoli Ma (University of Wisconsin-Milwaukee)

Zeolitic imidazolate framework (ZIF) membranes have demonstrated molecular sieving capability for the separation of similarly sized molecules. However, the low reproducibility in membrane synthesis remains a challenge for their scale-up for industrial application. Here, we report a vapor-phase seeding approach that combines atomic layer deposition (ALD) and ligand vapor treatment for the synthesis of high quality ZIF membranes with good reproducibility. In this method, ALD was used to deposit a uniform and conformal ZnO coating on the surface of support, followed by treatment using 2-methylimidazole vapor to convert ZnO to ZIF-8. A subsequent hydrothermal synthesis was employed to grow the seed layer into a continuous membrane. The resultant ZIF-8 membranes exhibited propylene permeances above 1×10-8 mol Pa−1 m−2 s−1and propylene/propane selectivities higher than 50. In addition, the membranes were found to be effective in separating 1,3-butadiene from a multicomponent C4 hydrocarbon mixture. Attempts to make ZIF membranes on polymeric supports via the vapor-phase seeding approach will also be discussed.

Abstracts – Oral Presentations, Monday

Presenting authors are indicated by an asterisk*. See the most up-to-date version of the program book with full abstracts at **www.membranes.org/nams~2022/**

Oral 6 – ORGANIC SOLVENT SEPARATIONS II

2:00 pm (Room Palm AD, Monday)

Oral 31 - Interfacial Polymerization with Hydrophobic Amines Provides Ultrathin Polyamide Nanofilms for Hydrocarbon Separation

Neel Rangnekar (ExxonMobil)*; siyao li (Imperial College London); Ruijiao Dong (Imperial College London); Valentina Elena Musteata (King Abdullah University of Science and Technology (KAUST)); Jihoon Kim (Queen Mary University of London); JR Johnson (Exxon Mobil); Bennett Marshall (ExxonMobil); Stefan Chisca (KAUST); Jia Xu (Imperial College London); Scott Hoy (ExxonMobil Research and Engineering); Benjamin McCool (ExxonMobil Research and Engineering); Suzana Nunes (King Abdullah University of Science and Technology (KAUST)); Zhiwei Jiang (Queen Mary University of London); Andrew Livingston (Queen Mary University of London)

Hydrocarbon separation is currently dominated by energy-intensive distillation, which relies on liquid-vapour phase changes. Liquid-phase membrane technology offers an energy-efficient alternative, but requires membranes resistant to organic solvents, with high permeance, and which are capable of enriching permeate streams in selected crude oil molecules. Herein, we synthesized hydrophobic amines, and used these to fabricate hydrophobic polyamide nanofilms by interfacial polymerization. These nanofilms provide >100 times faster transport of non-polar liquids than conventional hydrophilic polyamide nanofilms. In fractionation of light crude oil, manipulation of the film thickness down to ~10 nanometers achieves an order of magnitude higher permeance than membranes made from polymers of intrinsic microporosity, yet with comparable size- and class- based separation. This high permeance significantly reduces membrane area requirements, expanding the potential for utilization of ultrathin nanofilm composite membranes in crude oil fractionation.

2:30 pm (Room Palm AD, Monday)

Oral 32 - A Smart and Responsive Crystalline Porous Organic Cage Membrane with Switchable Pore Apertures for Graded Molecular Sieving

Zhiwei Jiang (Queen Mary University of London)*; Ai He (University of Liverpool); Hadeel Hussain (Diamond Light Source); Jonathan Rawle (Diamond Light Source); Michael Briggs (University of Liverpool); Yue Wu (University of Liverpool); Marc Little (University of Liverpool); Andrew Livingston (Queen Mary University of London); Andrew Cooper (University of Liverpool)

Membranes with high selectivity offer an attractive route to molecular separations, where technologies such as distillation and chromatography are energy-intensive. However, it remains challenging to fine-tune the structure and porosity in membranes, particularly to separate molecules of similar size. Here, we report a process for producing composite membranes that comprise crystalline porous organic cage films fabricated by interfacial synthesis on a polyacrylonitrile support. These membranes exhibit ultrafast solvent permeance and high rejection of organic dyes with molecular weights over 600 g·mol-1. The crystalline cage film is dynamic and its pore aperture can be switched in methanol to generate larger pores that provide increased methanol permeance and higher molecular weight cut-offs (1400 g·mol-1). By varying the water/methanol ratio, the film can be switched between two phases that have different selectivities, such that a single, 'smart' crystalline membrane can perform graded molecular sieving. We exemplify this by separating three organic dyes in a single-stage, single-membrane process.(1)

Reference:

1.He, A. et al. A smart and responsive crystalline porous organic cage membrane with switchable pore apertures for graded molecular sieving. Nat Mater (2022).

3:00 pm (Room Palm AD, Monday)

Oral 33 - Solution Processable Polytriazoles from Spirocyclic Monomers for Membrane-based Hydrocarbon Separations

Nicholas C Bruno (Georgia Institute of Technology)*; M.G. Finn (Georgia Institute of Technology); Ryan Lively (Georgia Tech); Ronita Mathias (Georgia Institute of Technology)

The thermal distillation of crude oil mixtures is an energy intensive process, accounting for nearly 1% of global energy consumption. Membrane-based separations are an appealing alternative or tandem process due to intrinsic energy efficiency advantages relative to distillation. We developed a family of spirocyclic polytriazoles from structurally-diverse monomers for membrane applications. The resulting polymers were prepared by a convenient step-growth method providing very fast reaction rates, high molecular weights and solubilities in common organic solvents, and non-interconnected microporosity. As membranes, these materials demonstrated successful fractionation of whole Arabian light crude oil and atmospheric tower bottoms feeds, significantly enriching the low-boiling components of the mixtures as well as removing trace heteroatom and metal impurities, providing opportunities to reduce the energy cost of crude oil distillation with tandem membrane processes. Membrane-based molecular separations under these demanding conditions is made possible by high thermal stability and a moderate level of dynamic chain mobility leading to transient interconnections between micropores.

3:30 pm (Room Palm AD, Monday)

Oral 34 - Organometallic Hybrid Nanofilms with Highly Interconnected Pore Architecture for Ultrafast Solvent Transport with Precise Molecular Separation

Bratin Sengupta (University at Buffalo)*; Qiaobei Dong (Gas Technology Institute); Ji Jiang (Rensselaer Polytechnic Institute); Dinesh Behera (University at Buffalo); Fanglei Zhou (Membrane Technology and Research); Miao Yu (University at Buffalo)

Organic solvent nanofiltration (OSN) membranes with high solvent permeance and rigid, precisely controlled nanopores at Angstrom-scale are greatly needed to achieve effective molecular separation at low energy consumption. In order to obtain ultrahigh permeance, the most common route to pursue is to reduce the membrane thickness and thus shorten the effective molecular transport path; membranes as thin as one-atom thickness have been fabricated. In addition to membrane thickness, another usually overlooked factor, the pore interconnectivity described by the ratio of porosity and tortuosity (ε/τ), might also greatly influence the transport of molecules through the membrane. In this work, using a facile self-limiting interfacial reaction between a metallic precursor and an organic precursor, we fabricated a novel carbon-doped titania membrane that exhibits very high pore interconnectivity, as indicated by the high ε/τ; as a result, methanol permeance as high as 267 L m-2 h-1 bar-1 at room temperature was obtained, which is 2.5 times higher than the reported membranes with similar MWCO. Precise tuning of molecular weight cut-off (MWCO) in the range of 200-1,000 Da and with a step change of approximately 100 Da was realized by adjusting membrane synthesis conditions and calcination conditions. These membranes also have rigid nanopores that are stable in various organic solvents and at elevated temperatures, ensuring stable molecular separation under challenging conditions.

4:00 pm (Room Palm AD, Monday)

Oral 35 - Effect of xylene activity and crystal orientation of MFI- zeolite membranes on separation of xylene isomers

Fateme Banihashemi (Arizona State University); Jerry Lin (Arizona State University)*

Separation of xylene isomers by current technologies is highly energy intensive due to the similar physicochemical properties. Membrane based separation technology is a potential low-energy alternative. The present study examines the vapor permeation of xylene isomers through MFI zeolite membranes at various xylene activities. Using randomly and b-oriented MFI zeolite membranes synthesized in our lab, the influences of xylene activity and membrane microstructure on xylene vapor and liquid permeation/separation performance were s¬-ystematically investigated. The permeance for both p-xylene and o-xylene decreases with increasing xylene activity, but the decrease is more pronounced for p-xylene resulting in a reduction of p-xylene/o-xylene selectivity with increasing xylene activity. The lowest separation factor is observed when operated in the pervaporation mode giving p-/o-xylene separation factor of about 2 for MFI zeolite membranes of both microstructures. The b-oriented MFI zeolite membrane fabricated exhibits higher p-xylene selectivity than the randomly oriented membrane due to the presence of less intercrystalline gaps in the b-oriented zeolite layer, and the lesser decrease in p-xylene selectivity for the b-oriented membrane was confirmed with increasing xylene activity using vaper permeation mode but exhibit a similarly poor separation factor while operated in pervaporation mode. The results call for more research to improve the industrially important xylene activity dependence of xylene separation performance for MFI zeolite membranes.

[1] J. Mater. Chem. A, 2021, 9, 23793–23801.

4:30 pm (Room Palm AD, Monday)

Oral 36 - Data-driven development of polymer membrane materials for complex mixture separations

Youngjoo Lee (Georgia Tech)*; Lihua Chen (Georgia Tech); Rampi Ramprasad (Georgia Tech); Ryan Lively (Georgia Tech)

Separating liquid-phase complex mixtures relevant to chemical manufacturing and crude oil refining is a significant outstanding challenge in the field of membrane science. The development of new polymeric materials can provide a feasible solution to this challenge due to their variable separation capability, scalability, and durability. However, the time and investment to materialize and experimentally test for complex mixture separations are such high, a high cost which inhibits rapid material development in this emerging area. In this regard, a systematic predictive capability to predict the fractionation of complex liquid feeds via any arbitrary polymer membranes would expeditiously facilitate new polymer membrane development and separation process design. In this work, we have developed a data-driven prediction model that is capable of computing guest diffusivities as well as solubilities in arbitrary linear polymer membranes via machine learning algorithms. Those transport properties are used in a Maxwell-Stefan transport model to calculate the flux of each molecule in a complex mixture. The machine learning algorithms and transport modeling framework enable us to predict the permeations of any complex mixtures through any linear polymer membranes. The whole prediction process requires no experimental pre-inspection. In this work, the model is used to predict increasingly complex multicomponent liquid permeation beginning with binary solvent-solvent separations and ending with the fractionation of a 60,000 molecule crude oil via new-to-the-world polymers for validation. The predicted composition of permeate from crude oil separation shows 10 % error with the experimentally obtained one. This work is expected to open new opportunities to rapidly accelerate new material development and membrane process design.

Oral 7 – CARBON CAPTURE II

2:00 pm (Room Palm CF, Monday)

Oral 37 - Direct Ocean Capture Using Membrane Contactors

Austin R Lieber (University of Pittsburgh)*; Joanna Rivero (University of Pittsburgh); Donna Hildebrandt (University Of Pittsburgh); chris snodgrass (University of Pittsburgh); William Gamble (University Of Pittsburgh); Zoe Neal (University Of Pittsburgh); Shanna Davidson (University Of Pittsburgh); Huda Usman (University Of Pittsburgh); Tagbo Niepa (University Of Pittsburgh); Katherine Hornbostel (University of Pittsburgh)

Direct ocean carbon capture is the efficient removal of CO_2 from ocean water and requires designing, developing, and evaluating new and existing membrane systems with maximum surface area and capture capacity. Hollow fiber membrane (HFM) and microencapsulated aqueous basic solvent contactors are parametrically studied through experimentation and 1D model simulation before conducting a preliminary techno-economic assessment (TEA). In each of these contactors, CO_2 diffuses from the seawater through a gas-permeable, polymeric membrane made of polypropylene (PP) or polydimethylsiloxane (PDMS) into aqueous sodium hydroxide (NaOH) or sodium carbonate (Na_2CO_3) CO_2 solvent. The HFM studies suggest that system configuration as co-flow or counterflow is insignificant for CO_2 separation performance as the solvent is never fully saturated down the length of the fiber. HFM transmembrane CO_2 flux improves most when the solvent to seawater flow ratio is adjusted from 1:4 to 4:1, where the former achieves a flux range of 2.3-5.6×10^(-9) kg m^(-2) s^(-1) while the latter achieves a flux range of 2.6-5.8×10^(-9) kg m^(-2) s^(-1). The microencapsulated solvent studies in a fluidized bed suggest that a breakthrough time between 10^{-20} hrs with minimal pressure drop occurs when the capsule diameter is between 400-600μm, bed porosity is between 0.5-0.7, and flow velocity is between 0.15-0.3 m s^(-1). Initial results of the TEA for both membrane contactors indicate a need for membrane surface treatments that lower the local seawater pH; where the HFM and microencapsulated solvent contactor capture costs, at pH=8.05, are $12,335 and $5,479 per tonne CO_2, respectively, compared to their capture costs, at pH=5.00, of $137 and $176 per tonne CO_2, respectively. These efforts illustrate the potential for two novel negative emissions membrane contactor technologies.

2:30 pm (Room Palm CF, Monday)

Oral 38 - Highly Permeable Substrates Prepared by Vapor-Induced Phase Separation for CO_2-Selective Composite Membranes

Ruizhi Pang (The Ohio State University)*; Yutong Yang (The Ohio State University); Yang Han (The Ohio State University); Kai Chen (The Ohio State University); Winston Ho (OSU)

A substrate with bicontinuous structure has been attractive due to its high surface porosity and well-interconnected bulk pore structure, which significantly decreases the transport resistance of thin-film composite (TFC) membrane in CO_2 capture. In this study, polyethersulfone (PES) substrate with bicontinuous structure was fabricated by vapor-induced phase separation. During the preparation of PES casting solution, 2-pyrrolidone (2-PD) was used as the solvent, and H_2O was used as phase separation modulator. The ternary phase diagrams were evaluated to clarify the phase separation mechanism. The preparation conditions including water concentration in the casting solution and water vapor exposure duration were optimized. The CO_2 permeance of the optimized substrate was 293,507 GPU, which was 13.3 times of the baseline substrate. The CO_2 permeance of the TFC membrane fabricated on the optimized substrate was 964 GPU at 57°C. Compared to the TFC membrane fabricated on the baseline substrate, there was an increase of 104 GPU while at the same time, the CO_2/N_2 selectivity of the TFC membrane was maintained at 159. The permeance improvement of the TFC membrane by using the optimized substrate was due to both reduced substrate and lateral diffusion resistances, which were elucidated well by the resistance-in-series model. Finally, a 100′ long optimized PES substrate with a width of 21" was prepared using a continuous roll-to-roll casting machine to demonstrate the scalability.

3:00 pm (Room Palm CF, Monday)

Oral 39 - Techno-economic assessment of postcombustion capture processes using high-performance nanoporous graphene membranes

Marina Micari (EPFL)*; Kumar Varoon Agrawal (Ecole polytechnique fédérale de Lausanne)

Membrane processes are emerging as an alternative to absorption for postcombustion CO_2 capture. Membranes based on nanoporous graphene selective layer (NSLG) have shown high promise in this regard with extremely large CO_2 permeance (10000 GPU) and attractive CO_2/N_2 selectivity (30) [1,2]. We report a techno-economic model and assess the most performing process configuration for CO_2 capture from wet flue gas (XCO_2 of 13.5%).

We analyze the role of feed and permeate pressure on energy consumption and capture penalty, by comparing scenarios with compression in the feed channel, vacuum in the permeate channel or a combination of these. When the permeate is under vacuum, the energy consumption corresponding to a certain pressure ratio is lower than when feed is compressed, but the membrane area increases. However, the high permeance of graphene membranes uniquely makes this scenario more competitive than feed compression in most cases (large range of pressure ratio and membrane cost) [3].

Optimizing the double-stage process configuration, we find a minimum capture penalty of 36.8 $/ton with permeate pressures of 0.05 and 0.1 bar in the first and in the second stage, with membrane module cost of 500 $/m^2. The first stage covers almost 80% of the cost and is primarily responsible for CO_2 recovery, thus requires a much larger membrane area than the second stage. Therefore, the capture penalty can be further reduced by combining feed compression and permeate vacuum in the first stage, to reduce the membrane area. The optimized process presents feed and permeate pressures of 3 and 0.15 bar in the first stage and of 1 and 0.2 bar in the second stage and the capture penalty is 32.9 $/ton. These results show that NSLG membrane-based CO_2 capture is competitive with the optimized polymeric membrane-based systems.

[1] G. He et al., Energy Environ. Sci. 12 (2019) 3305-3312

[2] S. Huang et al., Sci. Adv. 7 (2021) eabf0116

[3] M. Micari et al., J. Membr. Sci. 624 (2021) 119103

3:30 pm (Room Palm CF, Monday)

Oral 40 - Critical aspects of high-pressure CO_2-induced plasticization in polyimide membranes

Menno Houben (Eindhoven University of Technology)*; Romy van Geijn (Eindhoven University of Technology); Joey Kloos (Eindhoven University of Technology); Machiel van Essen (Eindhoven University of Technology); Zandrie Borneman (Eindhoven University of Technology); Kitty Nijmeijer (Eijndhoven)

CO_2-induced plasticization is a major challenge for polymeric membranes for gas separation. These membranes often lose much of their selectivity, especially at high CO_2 pressures, due to plasticization resulting in poor membrane performances. CO_2 is in its supercritical (sc) state above 74bar and 31°C, which has significantly different properties compared to gaseous CO_2. Sc-CO_2 is often utilized as an extraction agent in the food industry or in enhanced methane recovery processes. The effect of the CO_2 fluid properties on the membrane performance and plasticization behavior is not much studied, while it has a large impact in high-pressure applications. In this presentation we explain the effect of high CO_2 pressures (up to 120bar and 55°C) on the plasticization of polyimide membranes.

The transition from gaseous-like to liquid-like sc-CO_2 was found to have the most influence on the CO_2 density and thus also influences the CO_2 sorption and permeability. The CO_2 sorption is closely related to the CO_2 density and determined the extent of plasticization, but showed no time-dependent behavior or hysteresis in the liquid-like region. In contrast, the CO_2 permeability strongly increased with increasing feed pressure and showed significant time-dependent behavior even in the liquid-like region. However, the extent of plasticization was found to be independent of the applied feed pressure in the liquid-like region. Moreover, the exposure to liquid-like sc-CO_2 changed the permeation behavior at all subsequent pressures due to severe plasticization and low chain relaxation rates. These results proof that the CO_2 fluid properties, exposure time and permeation history are critical aspects for plasticization at high CO_2 pressures. Finally, suppression of plasticization was achieved using three conventional methods (blending, thermal and crosslinking treatments), which showed that a strong interchain rigidity is required to suppress plasticization at liquid-like conditions.

4:00 pm (Room Palm CF, Monday)

Oral 41 - In situ synergistic growth of crystalline and polymer-incorporated amorphous ZIF-8 in polybenzimidazole achieving hierarchical nanostructures for H_2/CO_2 separation

LEIQING HU (University At Buffalo)*; Vinh Bui (University At Buffalo); Haiqing Lin (The State University Of New York At Buffalo)

Polybenzimidazole (PBI) with a strong size-sieving ability and excellent thermal stability is a leading membrane material for H_2/CO_2 separation at elevated temperatures. For example, commercial m-PBI exhibited H_2 permeability of 9.0 Barrer and H_2/CO_2 selectivity of 14 at 100 oC. Incorporating porous fillers in polymers to form mixed matrix materials (MMMs) is an effective way to improve their performance. However, most porous fillers have lower H_2/CO_2 selectivity than PBI, leading to higher H_2 permeability but lower H_2/CO_2 selectivity in the MMMs. In this study, we demonstrate the in-situ synergistic growth of crystalline zeolite imidazole framework-8 (ZIF-8) and polymer-incorporated aZIF-8 MMMs (CPAMs). The formed porous crystalline ZIF-8 (cZIF-8) increases gas permeability. More importantly, PBI contains benzimidazole groups on the backbones (similar to the precursor for ZIF-8, i.e., 2-methylimidazole) and induces the formation of amorphous ZIF-8, enhancing interfacial compatibility and tightening the PBI nanostructures. The structures of CPAMs were thoroughly characterized using FTIR, WXRD, TGA, and SEM. The crystalline ZIF-8 NPs have sizes of 0.7 - 1 μm, and amorphous ZIF structures distribute evenly in the films. For example, the formation of 15 mass% ZIF-8 in PBI improves H_2 permeability and H_2/CO_2 selectivity by ≈100% at 35 ℃. The effects of feed gas composition, water vapor, temperature, and operation time on the separation properties were investigated. The CPAM shows stable H_2/CO_2 separation performance (H_2 permeability of 42 Barrer and H_2/CO_2 selectivity of 28) at 200 oC, breaking the permeability/selectivity tradeoff, surpassing the Robeson upper bound and demonstrating its potential for precombustion carbon capture. This work, for the first time, unveils a new platform of MMMs comprising functional polymer-incorporated amorphous ZIFs with hierarchical nanostructures for various applications.

4:30 pm (Room Palm CF, Monday)

Oral 42 - Mitigated carrier saturation of facilitated transport membranes for decarbonizing dilute CO_2 sources

Yang Han (The Ohio State University)*; Winston Ho (OSU)

Although carbon capture and storage (CCS) from large stationary sources is a leading strategy to reduce the carbon emissions, a carbon-neutral scenario in the energy sector cannot be achieved by the current CCS technologies since 90% CO_2 recovery is commonly assumed in the design. By taking coal-derived flue gas as an example, a 90% recovery through a primary capture system results in a residual flue gas containing 1–2% CO_2. If not captured, the residual emissions must be mitigated via direct air capture (DAC), which is usually hampered by a substantial cost. The focus of this presentation is to address the membrane carbon capture from dilute, yet stationary sources in order to achieve an overall >90% CO_2 recovery and thereby alleviate burdens upon DAC. To this end, the CO_2/N_2 separation performances of facilitated transport membranes (FTMs) containing aminoacid salts as mobile carriers were characterized with a CO_2 concentration of 0.05–20%. At a reduced CO_2 partial pressure, the carrier saturation in the FTMs was mitigated, which enhanced both the CO_2 permeance and CO_2/N_2 selectivity. The best FTM containing 2-(1-piperazinyl)ethylamine sarcosinate exhibited an uprising CO_2 permeance from 1968 to 3822 GPU and an improved CO_2/N_2 selectivity from 249 to 472 with reducing CO_2 content from 1% to 0.1%. The feasibility of this FTM is exemplified by designing a two-stage enriching membrane cascade to further remove 90% of the CO_2 in a residual coal flue gas containing 1.75% CO_2. Techno-economic analysis indicates a low capture cost of $83.8/tonne. The marginal costs beyond 90% capture are also evaluated for a variety of residual flue gases, indicating that the FTM-based capture from the coal or cement plant residual flue gas is more cost effective than DAC.

Oral 8 – EMERGING MATERIALS FOR LIQUID SEPARATIONS II

2:00 pm (Room Palm BE, Monday)

Oral 43 - Bioinspired cross-linked iron doped reduced graphene oxide (rGO) membrane for ultrafiltration of organic micropollutants

Yichen Wu (McMaster University); Nan Zhang (McMaster University); Gavin Yuen (McMaster University); Charles-François de Lannoy (McMaster)*

Reduce graphene oxide (rGO) has been shown to enhance the adsorptive ability of membranes for a range of organic contaminants. However, rGO surface-modified membranes demonstrate very low water fluxes and poor stability due to the high rGO interlayer packing density and swelling of rGO nanosheets. We have synthesized a crosslinked rGO nanocomposite membrane, exhibiting high water permeance and greatly improved dimensional stability by pursing a unique and simple inter-sheet chemistry: branched polyethylenimine (HPEI). These membranes were formed by a layer of iron doped rGO (rGO – Fe) cross-linked with branched polyethylenimine (HPEI) vacuum deposited onto a polydopamine (PDA) surface-modified polyethersulfone (PES) ultrafiltration membrane. The HPEI cross-linking of iron-doped rGO was accomplished by 1) activating the carbonyl groups of the iron nanoparticles anchored on graphene oxide (GO – Fe) by standard EDC-NHS reaction chemistry, 2) cross-linking GO – Fe through the primary amines of HPEI, and 3) thermal reduction of crosslinked GO – Fe by $NaBH_4$. The resulting rGO nanocomposite membrane was more stable in water than previously reported rGO membranes, showing an ability to remain surface stable at different pHs for over 20 days. HPEI crosslinking and in-situ growth of nano-sized Fe on rGO nanosheets expanded the rGO interlayer spacing allowing higher water permeance. The optimized membrane achieved a water permeance of 39.8 , which was 2-fold greater than most reported GO or rGO-based membranes. Adsorption, size exclusion, and pH-dependent electrostatic interactions controlled the separation of charged solutes, resulting in removals above 75.8%, 81.6% and 95.7% respectively for p-nitrophenol (PNP, 6.5 mg/L) and methyl orange (MO, 20 mg/L0) and methylene blue (MB, 20 mg/L) during dead-end ultrafiltration. This study provides an enhanced material framework for rGO nanocomposite membranes for micropollutant removal.

2:30 pm (Room Palm BE, Monday)

Oral 44 - Functional nanofibrous membranes modified with β-cyclodextrins for improved surface characteristics and filtration performance

Saranya Rameshkumar (Trinity College Dublin)*; Ramesh Babu Padamati (Trinity College Dublin)

Polymeric nanofibrous membrane modified with hydroxypropyl β-cyclodextrins (HP β-CDs) are fabricated and characterised for selectively separating out organic solutes of bioeconomic importance. Facile modification of nanofibers was carried out by means of electrospraying technique to achieve active functional layer on the nanoporous membrane. Incorporation of HP β-CDs by in-situ crosslinking has resulted in nanofibers with narrow average fiber diameter of as low as 200 nm. The electro-fluidic deposition of HP β-CDs has formed an interconnected layer over the nanofiber surface and subsequent graft polymerisation was done post-electrospraying to develop stable functionalised membranes. SEM morphology and surface topography have revealed the presence of HP β-CDs containing surface layer which has offered reduced pore size in the range between about 150-200 nm as desirable for separation of low molecular weight organic solutes. HP β-CDs cross-linking has offered surface hydrophilicity as confirmed by the reduction in contact angle to 15.8° and 64.2° respectively for in-situ crosslinked and surface functionalised polyethersulfone (PES) nanofibers. Further characterisation analysis such as AFM, Raman have also been performed to understand its surface topography and functionality. Surface modification has also increased the membrane flux rate despite having lower pore size, which implies that both solution-diffusion property and hydrophilicity of toroidal structured HP β-CDs were offering improved membrane water permeability. Evaluation of functionalised PES nanofibers has shown nearly improved lactose efficiency with reduced permeability trade-off confirming HP β-CDs as one of the excellent moieties for developing surface functional layer for realising superior filtration properties towards organic solute separation.

3:00 pm (Room Palm BE, Monday)

Oral 45 - Cross-Linkable/ Self-Assembling Terpolymer Membranes for Brackish Water Desalination

Samuel J Lounder (Tufts)*; Ayse Asatekin (Tufts)

Water filtration membranes with advanced ion selectivity and fouling resistance are urgently needed for applications such as desalination, water softening, and selective ion separations. This work introduces cross-linkable/ charged zwitterionic copolymers (x-CZACs) to develop the first self-assembled copolymer membranes capable of brackish water desalination. This new addition to the cross-linkable zwitterionic copolymer family combines cross-linkable/ hydrophobic, zwitterionic, and net-charged ionic monomer building blocks. We show that x-CZACs self-assemble to form a bicontinuous network of water-permeable, hydrophilic nanochannels surrounded by a cross-linkable/ hydrophobic nanodomain. As with previous cross-linkable zwitterionic copolymer membranes, polymer cross-linking is performed to tune self-assembled nanochannel diameter to <1.0 nm. A key difference, however, is that our choice in hydrophilic chemistry led to the formation of net-charged nanochannels for increased salt rejection by Donnan exclusion. Through this intentional adaptation to hydrophilic chemistry, x-CZAC membranes achieve >97% salt retention for brackish-salinity feed solutions (2,000-6,000 ppm NaCl). Salt permeation is understood by Maxwell-Stefan transport modeling and salt sorption measurements, indicating that Donnan exclusion plays a crucial role in increasing salt retention. This work illustrates that cross-linkable/ self-assembling copolymers offer a scalable route for developing membranes with customized ion separation performance, all dependent on the choice in hydrophilic copolymer chemistry.

3:30 pm (Room Palm BE, Monday)

Oral 46 - Including non-binding ‘spacer’ monomers in polyprotic polymeric ligands impacts ligand-ion affinity for lanthanum

Priyanka Suresh (Case Western reserve University)*; Christine E Duval (Case Western Reserve University); Ming Yu (Case Western Reserve University); Amy Che (Case Western Reserve University); Kevin Pataroque (Case Western Reserve University); Dylan Kulbacki (Case Western Reserve University)

Membrane adsorbers (MAs) are a useful tool for selectively capturing heavy metals like lanthanides and actinides from water. Ion-selectivity is imparted by coating ultra- or microfiltration membranes with ligand-containing polymer brushes or networks. Most commonly, these ligand-containing polymers are homogeneous (made from one monomer). Herein, we show that membranes coated with heterogenous polymer brushes (i.e. copolymers of 'spacer' monomers and ligand-bearing monomers) exhibit distinctly different ion-interactions (affinity, capacity, and kinetics) than their homogeneous counterparts. Thus, introducing co-monomers is an effective strategy for tuning the separation performance of MAs.

In this contribution, the ligand-bearing monomer is an amphoteric phosphonate ($O\text{-}PO(OH)_2$), ethylene glycol methacrylate phosphate (EGMP) and the model ion is La^{3+}. Monomers in this study are EGMP (La^{3+} binding), 2-hydroxyethyl methacrylate (HEMA: does not bind La^{3+}, H-bond donor), and butyl methacrylate (BuMa: does not bind La^{3+}, does not H-bond). Membranes were coated with varying polymer compositions using thermal polymerization and characterized using ATR-FTIR, NMR, and contact angle goniometry. Copolymer composition was measured by elemental analysis (P, N, O). The degree of grafting was calculated gravimetrically and was consistent (~0.2 g/g) for all copolymer coatings: poly(EGMP), poly(EGMP-co-HEMA) and poly(EGMP-co-BuMa). Binding capacities for La^{3+} were estimated using the Langmuir isotherm and normalized to the mass of EGMP per membrane, for comparison. The Langmuir constant, KL (L/mg)—an indicator of La^{3+} affinity, is markedly higher when a non-H bonding 'spacer' is introduced: poly(EGMP) = 0.02 < poly(EGMP-co-HEMA) = 0.04 < poly(EGMP-co-BuMa) = 0.69. A similar trend is seen for kinetic rate constants in which faster adsorption is observed for polymers with less H-bonding. These data indicate that intermolecular H-bonding has a negative effect on ion-adsorption.

4:00 pm (Room Palm BE, Monday)

Oral 47 - Enhancing the salt rejections of covalent organic framework membranes by creating dual-layer membrane structure

Miguel Jaimes (UWM)*; Zhiqin Qiang (University of Wisconsin-Milwaukee); Rahul Khandge (University of Wisconsin - Milwaukee); Ho Kuan Yu (University of Wisconsin - Milwaukee); Weiling Xia (University of Wisconsin - Milwaukee); Rachel Clark (UW - Milwaukee); Xiaoli Ma (University of Wisconsin-Milwaukee)

Covalent organic frameworks (COFs) are an emerging class of porous materials constructed from organic building blocks via strong covalent bonds. They possess a range of unique properties, including high crystallinity, ordered pore structure, tunable pore size and functionality, and excellent stability, making them promising materials for membrane applications. Most COF membranes reported in the literature exhibited low salt rejections due to their relatively large pore size. Here, we report the creation of dual layered structure in COF membranes to enhance their desalination performance. The COF membranes were synthesized on polymer support in a matter of seconds via the scalable interfacial polymerization method. Although the COF membranes formed appear to be amorphous, we have demonstrated that the salt rejection properties can be tuned by controlling the COF-COF interface in dual layer COF membranes. The misalignment between the distinct 2D COF structures has possibly narrowed the effective membrane pore size. As a result, the dual layered COF membranes exhibited higher salt rejections than single layer COF membranes.

4:30 pm (Room Palm BE, Monday)

Oral 48 - Polyampholyte copolymer self-assembly for fouling resistant and easily tunable membranes

Luca Mazzaferro (Tufts University)*; Samuel J Lounder (Tufts); Ayse Asatekin (Tufts)

Membrane separations are energy-efficient, simple, and scalable. Yet, their broader use is limited by the separation capabilities of membranes prepared by conventional methods, typically susceptible to fouling and with pour selectivity due to high pore size polydispersity. Self-assembly of functional polymeric materials is a powerful platform for designing membranes capable of new and more efficient separations. Previous researchers have developed a method for preparing such membranes by the spontaneous self-assembly of random amphiphilic copolymers, it provides a combination of scalability and good performance. Here we present a new class of self-assembled random amphiphilic copolymers with tunable selectivity and good fouling resistance based on polyampholyte random amphiphilic copolymers. Polyampholytes are copolymers with positively and negatively charged monomer subunits. In this study we demonstrate that polyampholyte random amphiphilic copolymers are highly fouling resistant, and that it is possible to tune salts and small organic molecules selectivities by controlling the ratio of positive and negative monomer subunits of our copolymers. We were able to vary the rejection of sodium sulfate from 5% to 93% with no significant change in pore size or fouling resistance. We also demonstrate that polyampholyte selective layers are a promising platform to create membranes with good performance and easy fabrication.

Oral 9 – MEMBRANE SYNTHESIS AND CASTING II

2:00 pm (Room Abbey North, Monday)

Oral 49 - Thin zeolitic-imidazolate framework ZIF-8 membranes supported on ZnO-deposited substrates for propylene/propane separation

Anil Ronte (Oklahoma State University)*; Phadindra Wagle (Oklahoma State University); Diako Mahmodi (Oklahoma State University); SHAILESH DANGWAL (Oklahoma State University); Elena Echeverria (Oklahoma State University); Dave Mcilroy (Oklahoma State University); Seok Jhin Kim (Oklahoma State University)

Zeolitic-imidazolate framework-8 (ZIF-8) membranes have gained attention due to their separation capability of propylene/propane. This study focuses on developing thin ZIF-8 membranes on ZnO-deposited supports using a vapor phase technique. The method included atomic layer deposition (ALD) of ZnO thin film on the Anodic Alumina Oxide (AAO) substrates and subsequent conversion from ZnO to ZIF-8 layer using vapor-phase 2-methyl imidazole (2-mIm) treatment. The method enabled solvent-free synthesis, which reduced the number of chemicals in the fabrication, and resulted in ultrathin (~350 nm) propylene selective ZIF-8 membranes. A series of characterization techniques such as X-ray diffraction, FT-IR spectroscopy, and electron microscopy were used to investigate the surface properties of the ZIF-8 membranes. The vapor-phase synthesized ZIF-8 membranes were tested for the binary propylene/propane gas transport properties to evaluate the quality. The resulting ZIF-8 membranes showed an average propylene/propane separation factor of 44 with high propylene permeance of 2.3×10^{-7} mol m^{-2} s^{-1} Pa^{-1}.

Keywords: ZIF-8 membrane, Atomic layer deposition, Propylene/propane separation, Vapor-phase synthesis.

2:30 pm (Room Abbey North, Monday)

Oral 50 - 3D Printing of Thin Film Composite Membranes for Nanofiltration and Beyond

Jeffrey R McCutcheon (University of Connecticut)*; Xin Qian (University of Connecticut); Mayur Ostwal (University of Connecticut); Edward Wazer (University of Connecticut); Noah Ferguson (University of Connecticut); Marianne Pemberton (University of Connecticut)

Making membranes through additive manufacturing (i.e. “3D printing”) has emerged as a popular field of study in membrane science over the past half-decade. Additive manufacturing confers a number of benefits when making membranes including customizability, thickness control, morphology control, and hierarchical design. Alternative manufacturing techniques also enable the processing of emergent materials into high performance membranes. This is important because while the number of new materials in membranes has increased dramatically, the options for processing those new materials has not.

Our work has identified a new approach to making membranes through additive manufacturing using electrospray deposition. This method enables the processing of classic and emerging polymer materials into thin film composite (TFC) membranes. Unlike current TFC membranes, we do not need to form these selective layers in-situ.

Our work demonstrates the ability to print membranes for a variety of separations using a number of emergent polymers, such as zwitterionic copolymer (nanofiltration), polyether epoxy (chlorine tolerant nanofiltration), and PIM-1 (gas separations). Traditionally, these polymers would be processed using a phase inversion process, leading to an integrated asymmetric membrane with relatively low permeance. Our work demonstrates the versatility of the method to make TFC membranes with different polymers and yield selective yet substantially higher permeance due to our ability to make thinner, defect free membranes.

We contextualize this work in the broader perspective of membrane manufacturing which has seen little innovation in the past 4 decades. Innovations in membrane manufacturing will enable the use of the increasing number of new membrane materials and could help realize the benefits promised by these materials.

3:00 pm (Room Abbey North, Monday)

Oral 51 - Advanced ionic polymers for high-performance gas separation membranes

Sudhir Ravula (University of Alabama)*; Jason Bara (University of Alabama)

Global warming, a primarily driven by human activities, particularly burning fossil fuels, which increases the gas levels (e.g., CO_2) and traps the heat in the earth's atmosphere and rises the earth's surface temperature. Removal of greenhouse gases (e.g., CO_2, CH_4) from flue gas has been extensively studied by membrane technologies for an efficient gas separation process. Among them, ionic polymers have been developed based on functionalized room temperature ionic liquids building blocks for selective gas properties. In general, ionic polymers are classified into two groups based on their association in their molecular structure. First, polymerized ionic liquids or poly(ionic liquid)s, poly(IL)s are ion-containing polymers, wherein the ionic moieties are pendant from the backbone. Secondly, ionenes are ionic polymers that contain ionic (typically cationic) moieties connected along the backbone of the polymeric chain. We have developed high-performance (HP) ionenes and poly(IL)s materials and studied their applicability for membrane-based gas separations. The synthesized ionic materials have been characterized using Fourier transform infrared spectroscopy, nuclear magnetic spectroscopy, gel permeation chromatography, thermogravimetric analysis, and differential scanning calorimetry to evaluate the purity and other associated properties. These ionic materials exhibit good thermal and mechanical properties and quantified their structure-property relationship by analyzing pure gas permeabilities and selectivities. Further, the thermophysical properties of these ionic materials are enhanced when non-covalent "free" ILs are incorporated between the polymeric chain. In this talk, we will emphasize the synthesis and characterization of ionic polymers and ionic composites, as well as the gas separation performance.

3:30 pm (Room Abbey North, Monday)

Oral 52 - Confined Thermally Induced Phase Separation under Micromolding – Fabrication of Patterned Membranes with Enhanced Pore Connectivity and High Salt-scaling Resistance

Shouhong Fan (University of Colorado at Boulder)*; Yifu Ding (Membrane Science, Engineering, and Technology (MAST) Center, Department of Mechanical Engineering, University of Colorado Boulder); Anthony Straub (University of Colorado Boulder); Kamalesh K Sirkar (New Jersey Institute of Technology); Adrienne Blevins (Materials Science and Engineering Program, University of Colorado Boulder); Duong T. Nguyen (University of Colorado Boulder); Jaylene Martinez (CU Boulder Mechanical Engineering); John Chau (New Jersey Institute of Technology)

It is challenging to recover freshwater from hypersaline solution with conventional reverse osmosis (RO). Membrane distillation (MD), as a combination of thermal distillation and membrane process, is a promising method to treat high salinity water with nearly 100 % salt/particle rejection. Nevertheless, MD process suffers from permeation reduction due to membrane fouling and temperature polarization in the boundary layers between bulk flows and membrane surface, which severally diminish the economic viability of MD system. Membranes having properly designed patterns on the surface have been experimentally demonstrated for fouling mitigation, local shear force and mass transfer enhancement for the liquid-based separation process. Ultimately, the presence of surface patterns has a great potential to mitigate the extent of membrane fouling and temperature polarization whereby improves MD efficiency. To date, the factors that controlled patterns formation on membrane surface and the impact of surface patterns on MD desalination performance have not yet been systematically studied.

In this presentation, we demonstrate a novel method, namely, lithographically templated, thermally induced phase separation (lt-TIPS) for the fabrication of a series robust patterned PVDF microporous membranes which have bio-inspired 'sharkskin' surface patterns and pore structures varying from dispersed spherulitic to interconnected cellular. We will present the material formulation of lt-TIPS process and the factors that affect patterning fidelity on membrane surface e.g., diluent types, molding pressure, molecular weight of polymer, template surface chemistry. The impact of pore structure and presence of surface patterns on membrane permeability and MD performance will also be discussed. In addition, the effect of pattern dimensions on the feed side heat transfer coefficient and the resistance to salt-scaling will be thoroughly compared.

4:00 pm (Room Abbey North, Monday)

Oral 53 - Sustainability in Membrane Separations: Integration of Eco-Friendly Materials and Scaled-Up Fabrication of Polymeric Membranes for Water Purification

David Lu (University of Kentucky)*; Kwangjun Jung (Georgia Institute of Technology); Ju Yong Shim (Georgia Institute of Technology); Alex Williams (University of Kentucky); Kassandra Opfar (University of Kentucky); Isabel Escobar (University of Kentucky); Tequila Harris (Georgia Institute of Technology)

Polymeric Membranes have become more prevalent in separation applications due to their desirable chemical and physical properties, including versatility in pore size and application modes. Moreover, polymeric membranes are considered advantageous due to the relative ease in fabrication, particularly via phase inversion methods. Despite the benefits, the materials commonly used in the fabrication process present a serious drawback to polymeric membranes. Traditional solvents (e.g., NMP, DMAc, DMF) are largely derived from petrochemicals and present hazards to human health and the environment; when used in phase inversion methods, these solvents can leach into fabrication wastewater and enter water systems if adequate treatment methods are not used. As such, recent regulations that the use of several traditional solvents have motivated the search for suitable, greener (i.e., having bioderived, biodegradable, nonhazardous, and/or reusable properties) alternatives.

Among potential alternatives, Rhodiasolv® PolarClean and gamma-valerolactone (GVL) are two green solvents of potential interest for polymeric membrane fabrication. Both solvents are biodegradable and nontoxic, as PolarClean is derived from the by-product of Nylon 6,6 fabrication and GVL is sourced from lignocellulose. When used together as a co-solvent mixture, resulting polysulfone (PSf) UF membranes fabricated at the lab-scale have exhibited performance properties competitive to those of UF membranes that use traditional solvents. Beyond the lab-scale, optimization of fabrication parameters was investigated for scaled-up preparation of PSf/PolarClean-GVL UF membranes via slot-die coating. A comparison of the membrane performance and morphological characteristics from both fabrication methods and a study on the influence of evaporation time during the fabrication process were conducted. The findings of the study indicate the feasibility of utilizing green solvents and scale-up to improve sustainability.

4:30 pm (Room Abbey North, Monday)

Oral 54 - Polyamide thickness and morphology features of thin-film composite membranes depend on the extent of restriction of the amine monomer supply during interfacial polymerization

Mikayla D Armstrong (University of North Carolina at Chapel Hill)*; Kasia Grzebyk (UNC Chapel Hill); Orlando Coronell (University of North Carolina at Chapel Hill)

The polyamide layer of thin-film composite membranes (TFCs) is formed by an interfacial polymerization (IP) reaction. The membrane community regards the IP kinetics as limited by the diffusion of the amine monomer through the forming polyamide layer. Polyamide layers formed by the conventional method are < 350 nm thick and have a basal layer with a ridge-and-valley structure comprised of voids and leaf-like features, some of which are flat and overlapping. This work aimed to better understand the effect of MPD supply conditions during IP on polyamide thickness and morphology features. We modified existing and fabricated new polyamide layers with varied MPD supply conditions (e.g., permeation rate, delivery method, concentration, IP time) and analyzed their resulting thicknesses and morphology features. Results showed that modified and new membranes had average film thicknesses ranging 155–1206 nm and 116–1282 nm, respectively, with five morphology features. We observed three features identified previously—a basal layer and a secondary layer with flat, overlapping features —and three new features: multi-layered void structures, void mega-nodules, and a void-less blanket layer. The combination of MPD supply conditions that restricted MPD supply the least resulted in the thickest polyamide layer, 1282 nm. Multi-layered void structures and void mega-nodules developed when reduced restrictions in the external MPD supply conditions (i.e., delivery method, concentration, and IP time) were paired with high and low MPD permeation rates, respectively. The void-less blanket layer seemed to form before multi-layered void structures and void mega-nodules. The conditions of formation of each of these morphology features are consistent with recent findings on void formation caused by CO_2 degassing and nanobubble confinement conditions. Our work furthers the understanding of the IP reaction and provides new insights for the fabrication and modification of TFCs.

Oral 10 – CATALYTIC AND RESPONSIVE MEMBRANES

2:00 pm (Room Abbey South, Monday)

Oral 55 - Synthesis of Reactive Membranes for Capture and Degradation of Water and Air Pollutants

Rollie G Mills (University of Kentucky)*; Dibakar Bhattacharyya (U. Kentucky); Matthew D Bernard (University of Kentucky); Ronald J. Vogler (University of Kentucky); Jacob Concolino (University of Kentucky); Todd Hastings (University of Kentucky); Yinai Wei (University of Kentucky); Thomas Dziubla (University of Kentucky); Kevin Baldridge (University of Kentucky)

Clean air and water have become prominent needs in recent years due to the COVID-19 virus pandemic and the widespread presence of toxic chemicals, such as polychlorinated biphenyls (PCBs), in water sources. Exposure to such compounds can result in viral infection, neurological problems, and birth defects in humans, thus source filtration and contaminant removal is becoming increasingly vital for health longevity. To combat this problem, novel responsive materials were synthesized to capture and degrade these pollutants from air and water sources. For the filtration of viral-sized aerosol particles, commercial microfiltration membranes were functionalized with a protease enzyme that can cleave the spike glycoprotein (SGP) of the SARS-CoV-2 virus, rendering it unable to infect its host. This synthesis process also incorporated poly(acrylic acid), which is known for its water-retaining properties, thus enhancing the longevity of enzyme activity on the membrane surface without the presence of significant hydration. For PCB treatment in industrial water sources, similar reactive membranes were synthesized using zero-valent iron-palladium nanoparticles. With these particles, the membranes can dechlorinate PCBs and other harmful chlorinated compounds via electron-transfer reactions, thus significantly lowering the toxicity of such pollutants. The synthesis of these membranes shows exciting advances in creating the next-generation of respiratory face masks and filtration systems for air and water remediation, respectively. This research is supported by the NSF-RAPID program and by NIEHS.

2:30 pm (Room Abbey South, Monday)

Oral 56 - Ionic Strength and pH Responsive UF and RO membranes

Yoram Cohen (Professor)*; Yian Chen (UCLA)

Membrane surface tethered poly(acrylic acid) (PAA) chains (size range of 96-145 nm) were synthesized onto a tight commercial polysulfone (PSf) ultrafiltration (UF) membrane (pore size of ~6.3 nm) via atmospheric pressure plasma-induced graft polymerization (APPIGP). The surface nano-structured (SNS) UF membrane had hydraulic resistance and molecular weight cutoff (MWCO) in the ranges of 0.74 - 2.29×1013 m-1 and 1.8 - 15 kDa, respectively, in response to stimuli of changing pH in the range of 3-11, and ionic strength variation in the range of 0-547 mM. The self-adaptive performance variation of the SNS-PAA-PSf membrane, due to pH and ionic strength changes, is attributed to conformational change of the tethered PAA chains (i.e., swell/collapse). When the tethered polymer brush layer is in an extended conformation, surface hydrophilicity was shown to increases. The above behavior is attributed to the change of water volume retained within the tethered polymer layer. When the tethered PAA chains collapse, both the intrinsic membrane resistance and MWCO decrease, thereby overcoming the typical membrane permeability/selectivity tradeoff.

Modification of a loose TFC-PA reverse osmosis (RO) membrane with tethered PAA chains increased the nominal salt (NaCl) rejection from 98.4% to 99.5% as the feedwater salt concentration increased from 2 to 32 g/L. The SNS-PAA-PA membrane was adequate as a BWRO membrane at feedwater NaCl concentration <~6 g/L, and as SWRO membrane at NaCl concentration >~6 g/L. The above result is significant as it demonstrates that t the SNS-PAA-PA membrane is adaptive whereby it can perform both the function of a BWRO membrane and also serve as SWRO membrane at increased salinity. The applicability of the SNS-PAA-PA as BWRO or SWRO membrane could be tailored by tuning the PAA brush layer synthesis conditions (plasma and graft polymerization).

3:00 pm (Room Abbey South, Monday)

Oral 57 - Catalyst-free water-gas shift reaction in ceramic-carbonate dual-phase membrane reactors at high pressures – Effect of Side Reaction

Oscar Ovalle (Arizona State University); Jerry Lin (Arizona State University)*

Water-gas-shift reaction (WGS) is critical in integrated gasification combined cycles (IGCC) power plants with CO_2 capture. In this work, WGS reaction with CO_2 separation was studied in tubular ceramic-carbonate dual-phase (CCDP) membrane reactors (MR) at high temperatures(HT) and pressures(HP). This study aims to provide more details on the WGS reaction in a CO_2-permeable CCDP MR and understand the effect of unwanted side reactions on MR performance. Samarium-doped ceria (SDC) powders were synthesized by the citrate-precursor method, and dead-end tubular supports were fabricated by the cold isostatic pressing (CIP) method and sintering process at 1200°C. Dense membrane reactors were prepared by infiltrating lithium/sodium molten carbonates (MC) into the support pores. The WGS reaction was performed without and with a high-temperature WGS catalyst (Co-Mo-Mg $(AlO_2)_2$). WGS reaction tests were performed with a feed gas mixture of $CO/CO_2/N_2/H_2O$ of a molar concentration of 45.7/13.1/41.3% (dry-base) and steam to carbon (S/C) ratios of 4, 7, and 14. The feed gas was introduced in the shell side of the MR. The experiments were run at space velocities (SV) between 150-3000 h-1, temperatures in 700-950 °C, total feed pressure between 1-15 bar, and sweep gas of helium at 50 ml/min and 1 bar. The results show that the free-catalyst MR can convert 92% of CO into CO_2 and H_2 and recover 29% of CO_2 at 850 °C, 7 bar, S/C ratio of 4, and SV of 150 h-1; CO conversion overcomes the equilibrium due to Le Chatelier's principle and the CO_2 removal from the reaction side. However, unwanted carbon deposition is also promoted under these conditions.

Experiments of WGS reaction in MR with catalyst showed minimum carbon formation (less than 4%) with CO conversion of 82% at high SV of 1000 h-1, 850 °C, and pressure lower than 7 bar. The results suggest that to enhance the CO conversion in WGS reaction with minimum carbon deposition, the CCDP MR should be operated with a catalyst at HP and HT.

3:30 pm (Room Abbey South, Monday)

Oral 58 - Catalytic Membranes and the Role of Biohydrogen for Reductive Chloro-Organics Detoxification from Water

Dibakar Bhattacharyya (U. Kentucky)*; Rollie Mills (U. Kentucky); Hongyi Wan (University of Kentucky); Michael J Detisch (University of Kentucky); Todd Hastings (University of Kentucky); Noah Meeks (Southern Services Co)

The incorporation of nanostructured catalytic materials (such as, Pd) in microfiltration type functionalized membranes or creating metal-polymer composite membranes can create immense value in the area of pollutant remediation from water. Reductive processes with Pd containing catalyst in membranes require dilute hydrogen source for reductive dehalogenation. If one can produce hydrogen through photobacteria (near IR light) by converting waste organic acids to hydrogen, then we can integrate Pd-containing membranes for reductive degradation of water pollutants. The presentation will include: (a) Hydrogen generation by photo-responsive bacteria (Rhodopseudomonas palustris). We used NIR light sources and optically resonant gold-silica nanoparticles to increase light utilization of the bacteria to convert waste organic acids, such as acetic and maleic acids, to hydrogen (RSC Adv, 2019; Int. J. Hydrogen Energy, 2021). We demonstrated > 2.5 fold increases in hydrogen production when grown under a NIR source compared to a broad-band light source at equal light intensity (130 W/sqm). The use of the of the biohydrogen with functionaized membranes (PVDF-PAA-Pd, and Fe/Pd) containing Pd particles showed over 90% conversion of PCB-1 to biphenyl, indicating excellent hydrogenation; (b) Use of composite membranes (porous Pd film on UF membrane for PCB degradation. Magnetron sputtering was used to deposit thin films of tantalum and MgPd alloy on top of polysulfone (PSf) membranes 100 nm thick (J. Mat. Res, 2021). A dealloying step in water removes the Mg component of the film creating a nanoporous (np) film of Pd. PCB's were dechlorinated while permeated in a water solution under pressurization with 5% H_2 gas (remainder Argon). Over 60% of PCB-1 was degraded in solution with a single pass (under 4 bar pressure) through the composite membrane at 40 LMH. This research has been supported by NIH-NIEHS-SRP, NSF-EAGER, and Southern Company.

4:00 pm (Room Abbey South, Monday)

Oral 59 - Catalytic membrane reactor for conversion of waste biomass to chemical intermediates

Ranil Wickramasinghe (University of Arkansas)*; Zhexi Zhu (University of Arkansas); Xianghong Qian (University of Arkansas)

Agriculture and related industries are responsible for up to 24% of global greenhouse gas emissions. Conversion of waste biomass to chemical intermediates that can be further processed to fuels and biobased products will help extend the carbon cycle and reduce greenhouse gas emissions. As an example, it is estimated that 95% of food waste ends up in landfills, which eventually releases methane and carbon dioxide and contributes to greenhouse gas emissions. However economic conversion of this biomass to fuels and chemicals remains elusive. Intensified unit operations are required to reduce the manufacturing cost. A patented catalytic membrane reactor has been developed to hydrolyze and dehydrate lignocellulosic biomass. The solid acid catalyst consisting of dual polymer chains is grafted from the surface of a ceramic membrane. The acidic polymeric chain, poly(styrene sulfonic acid) (PSSA) catalyzes biomass hydrolysis as well as dehydration. A neighboring poly(vinyl imidazolium chloride) (PIL) chain helps solubilize lignocellulosic biomass and enhance the catalytic activity.

Hydrolysis was conducted for crystalline cellulose and acid, base or steam pretreated corn stover samples in ionic liquids (IL) and mixtures of IL with water or γ-valerolactone (GVL). Near quantitative total reducing sugar (TRS) yields for cellulose hydrolysis as well as pretreated corn stover biomass were achieved at mild conditions and in less than 12 h. Other lignocellulosic biomass residues such as vegetables, rice straw, wheat straw and almond shells have also been tested.

The catalytic membrane reactor may be used to produce sugars as well as chemical intermediates such as 5-hydroxymethylfurfural (HMF) furfural or levulinic acid. The catalysts are stable and maintain high catalytic activity after repeated runs. The catalyst can be easily regenerated. Our results indicate that the catalytic membrane catalyzes the reaction and recovers the product one unit operation.

4:30 pm (Room Abbey South, Monday)

Oral 60 - Trends, Mechanisms, and Opportunities in using Electrically Responsive Membranes

Melissa J Larocque (McMaster University); Adi Gelb (McMaster University); David Latulippe (McMaster University); Charles-François de Lannoy (McMaster)*

A range of conductive materials have been used to develop electrically responsive membranes (ERMs) for use in various applications, including fouling mitigation, selectivity enhancement, electrochemical degradation of contaminants, and surface sensing capabilities. ERMs have been fabricated both as stand-alone porous materials and as composite materials supported on a conventional membrane. We performed a meta-analysis of composite ECMs from 111 studies based on their physical properties and specific applications to identify their material limitations, performance trends, and avenues for future work to guide the community. Two major challenges were identified: the wide variety of test conditions (98 unique pollutants across 193 different experiments) used, limiting inter-study comparisons, and a very limited body of ERM research that uses realistic feed sources (e.g., municipal wastewater). We propose a set of characterization and performance tests that researchers can use as standards in comparing ECM materials and configurations. We mapped surface conductivity against hydraulic permeability data from all studies with available data using Robeson-like analysis to illustrate the respective trade-offs that exist when using graphitic nanomaterials and conductive polymers. Metal-based composite ECMs were not subject to the same trade-offs and generally out-performed the conductivity of other ECMs. The conductivity/permeability plots developed provide a standardized approach for evaluating the performance of composite ECMs, as well as serving as a design guide for decisions aimed at achieving specific performance levels. We further identified the following key areas for future work to better demonstrate ECM capability in industrial applications: 1) quantitatively optimize the frequency, amplitude, and duty cycle of applied currents, 2) increase the scale of ECM fabrication, and 3) increase the time length of experiments to confirm long-term performance.

Abstracts – Oral Presentations, Tuesday

Presenting authors are indicated by an asterisk*. See the most up-to-date version of the program book with full abstracts at **www.membranes.org/nams~2022/**

Oral 11 – POLYMERIC AND MIXED-MATERIALS – GAS SEPARATION I

9:30 am (Room Palm AD, Tuesday)

Oral 61 - Unexpected size-controlled vapor sorption in glassy membranes exhibiting configurational free volume

William J. Box (University of Oklahoma)*; Zihan Huang (University of Notre Dame); Ruilan Guo (University of Notre Dame); Michele Galizia (University of Oklahoma)

Conventional glassy polymers do not offer the possibility of simultaneously maximizing sorption- and diffusion-selectivity to achieve highly selective separations. Polymers exhibiting iptycene units may help overcome this limitation. The internal free volume of these structures is not related to the molecular conformation, such as the excess free volume in conventional glassy polymers, but to the molecular configuration. Moreover, the size of the internal volume of iptycene units is extremely regular and comparable to the size of one single molecule, which makes iptycene-based polymers highly selective. However, the fundamental mechanism of small molecules transport in iptycene-based polymers is still poorly understood. The few published sorption and transport data in these materials refer only to light gases, with little or no information about the sorption and transport behavior of bulky vapors. The scope of this study is to shed fundamental light on the latter aspect. We recently discovered that, while in conventional glassy polymers penetrant diffusion is size-controlled and sorption is enthalpy-controlled, which may create a trade-off between sorption- and diffusion-selectivity, vapor diffusion and sorption in a series of triptycene-based TR polymers are both size-controlled, which makes easier to simultaneously tune sorption-and diffusion-selectivity to achieve highly selective separations. The fundamental sorption and transport behavior of a series of condensable vapors exhibiting different molecular size was experimentally and theoretically investigated to elucidate the origin of this fully size-controlled transport behavior, and new exciting opportunities to exploit this unique feature for vapor separation and organic solvents separation are discussed. Finally, the mechanism by which iptycene units regulate penetrant transport in these materials is elucidated by combining experimental characterization and molecular modeling.

10:00 am (Room Palm AD, Tuesday)

Oral 62 - Microporous Polymers with Tailored Cavities by Direct Fluorination for Small Gas Molecule Separations

Wonhee Lee (Georgia Institute of Technology)*; Yi Ren (Georgia Institute of Technology); Conrad J Roos (Georgia Institute of Technology); Ryan Lively (Georgia Tech); Jong Geun Seong (Hanyang University); Jongmyeong Lee (Hanyang University); So Young Lee (Hanyang University); Yuseong Do (Hanyang University); Joon Yong Bae (Hanyang University); Sun Ju Moon (Hanyang University); Chi Hoon Park (Hanyang University); Hyejin Jo (Hanyang University); Ju Sung Kim (Hanyang University); Young Moo Lee (Hanyang University); Kueir-Rarn Lee (Chung Yuan University); Wei-Song Hung (Chung Yuan University); Juin-Yih Lai (Chung Yuan University)

Size differences between penetrating molecules and membrane pores govern mass transports and size-sieving behaviors in membrane-based separations. Engineering the membrane pores at the molecular level is significant for improving membrane productivity and efficiency, especially for the applications requiring complex molecular separations, such as natural gas processing or hydrogen recovery. Moreover, the regulation of mass transport at interfaces endowing membranes with cascaded pore size distributions from their surface can allow permeating molecules to be separated without reducing throughput.

Herein, we report the decoration of microporous polymer membrane surfaces via direct fluorination. This process has been applied for nonporous polymer membranes but was first adopted for microporous materials, including thermally rearranged (TR) polymers and polymers with intrinsic microporosity (PIMs). We observed that the surface pore size in ready-made microporous membranes was effectively regulated via C-H to C-F substitution toward the interiors of the micropores. The fluorine acts as angstrom-scale apertures that can be controlled the molecular transport, reducing the surface pore radii from 3.36 Å down to ~2.48 Å at 200-800 nm surface depending on the direct fluorination time.

As a result, the microporous polymers with cascaded cavities showed high molecular separation efficiencies in many industrially interesting gas separation applications above the corresponding upper bounds of He/CH4, H_2/CH_4, CO_2/CH_4, N_2/CH_4, and H_2/CO_2. Remarkably, we achieved tens of fold enhancement in separation efficiency in small gases recoveries such as helium and hydrogen but maintained membrane productivities. We also realized this pore-reorganizing concept on hollow fiber membrane modules. The modules still present improvement in gas selectivity, where He/CH_4 selectivity of TR and PIM hollow fiber modules increased from 24 and 1.5 to 800 and 61, respectively, after direct fluorination.

10:30 am (Room Palm AD, Tuesday)

Oral 63 - PIMs at Nanoscale

Wojciech Ogieglo (KAUST)*; Ingo Pinnau (KAUST)

Polymers of intrinsic microporosity (PIMs) and other microporous amorphous materials, like carbon molecular sieves (CMS), are very promising to tackle challenging energy intensive separations (natural gas, olefin/paraffin, solvent purification etc.). However, fabricating them at thickness size scales relevant for practical membranes (several micrometers or less) brings about many challenges including defect control, physical and chemical stability, excessive substrate pore intrusion, and others.

Being out-of-equilibrium, amorphous, high free volume systems both PIMs and CMS additionally suffer from physical aging, or progressive loss of excess free volume in time which often adversely impacts the achievable molecular flux[1].

In this work, we will discuss our recent work on single- and mixed-penetrant sorption in thin and ultra-thin PIMs[2] and CMS[3] films using in-situ spectroscopic ellipsometry. We find that the degree to which nano-confinement affects the sorption and dilation (swelling) of thin polymer films seems to be related with the nature and distribution of their micropores. Non-microporous polymers (cellulose acetate) show comparatively less impact of nano-confinement than microporous ones (PIM-1 or PTMSP). Moreover, the presence of ultramicropores (sub-0.7 nm) in PIM-1 seems to somewhat inhibit drastic collapse of the free volume in comparison to a polymer that essentially lacks ultramicroporosity (PTMSP)[4]. These implications may serve to help develop guidelines for the design of more physical aging-resistant membrane polymers.

References:

[1] Xu, L. et al. Carbon 2014, 80, 155-166.

[2] Ogieglo, W. et al. Macromolecules 2020, 53 (20), 8765-8774.

[3] Ogieglo, W. et al. ACS Appl Mater Interfaces 2021, 13 (18), 21765-21774.

[4] Ghanem, B. S. et al. Adv Mater 2014, 26 (22), 3688-92.

11:00 am (Room Palm AD, Tuesday)

Oral 64 - Polyformamidine as Fixed-Site Carrier for CO_2/N_2 Separation Membranes

Jingying Hu (the Ohio State University)*; Yang Han (The Ohio State University); Winston Ho (OSU)

Amidine species was perceived as better alternatives for amine species in the facilitated transport membranes for having faster CO_2 reaction kinetics. The synthesis of poly(ethylene formamidine) (PFA) was carried out via the polycondensation of ethylenediamine and triethyl orthoformate catalyzed by acetic acid. The synthesized PFA possesses the highest amidine functional group density owing to the case using the shortest-chain diamine, ethylenediamine. Moreover, in comparison with mobile carriers, it can have a much higher boiling point, which practically eliminates the possibility of carrier volatility and leaching out. In order to obtain the preliminary gas permeation results, PFA (10 kDa MW) was blended with polyvinylalcohol (PVA, 1.1 MDa MW) and then coated onto nanoporous polyethersulfone (PES) substrates to form 170-nm thin membranes. Increasing the content of PFA from 36 wt.% to 66 wt.% in these membranes led to a substantial increase in membrane performance, especially for CO_2 permeance. Further increasing the PFA content to 83 wt.% still enhanced the CO_2 permeance but showed a moderate drop in CO_2/N_2 selectivity. Under an extremely high PFA content of 90 wt.%, the membrane showed a low CO_2 permeance of 1019 GPU but a high CO_2/N_2 selectivity of 236, indicating severe pore penetration. The membrane containing 83 wt.% PFA exhibited the best results with 2807 GPU CO_2 permeance and 72 CO_2/N_2 selectivity at 77°C. In view of the superior transport performance, it is necessary to increase the MW of PFA to make it capable of forming the membrane matrix alone to enable the incorporation of CO_2-reactive mobile carriers, thereby improving the overall CO_2 separation performance. By adopting the design of experiment technique, the synthesis conditions of PFA were improved in terms of the temperature profile, catalyst amount, and monomer concentrations, through which the MW was successfully increased to 3.3 MDa.

11:30 am (Room Palm AD, Tuesday)

Oral 65 - Scalable One-Step Fabrication of MOF-based Asymmetric Mixed-Matrix Membranes with Exceptionally High Propylene/Propane Separation Performance

Yinying Hua (Texas A&M University)*; Sunghwan Park (Johns Hopkins University); Hae-Kwon Jeong (Texas A&M University)

Due to their potential, mixed-matrix membranes (MMMs) have been extensively studied for the last three decades or so[1]. There are, however, no commercially available MMMs due to their engineering challenges for scalable MMM fabrication[2, 3]. Recently, our group developed a scalable mixed-matrix membranes fabrication strategy using in-situ filler formation approach, named PMMOF[4]. The PMMOF decouples the membrane formation step from the filler incorporation step using an in-situ filler formation in the polymer, enabling formation of MMMs not only with much improved gas separation performances[4] but also in a scalable geometry (i.e., asymmetric hollow fibers with submicron thick selective MMM skin layers)[5]. However, the PMMOF suffers from two major challenges: 1) multi-step involved that are relatively complicated and 2) limited choice of polymer matrices (i.e., polyimide-based polymers).

In this presentation, we plan to present a novel second-generation one-step MOF-based asymmetric MMM fabrication strategy, addressing the challenges of the first-generation PMMOF process. The new process, termed "PIMOF", enable rapid formation of ZIF-8-containing asymmetric MMMs. The as-prepared MMMs showed unprecedentedly high C_3H_6/C_3H_8 (C3) separation performance as compared to other propylene-selective MMMs reported so far with the C_3H_6 permeance of ~ 7.5 GPU and C_3H_6/C3H8 separation factor of ~ 107. It is worthy of noting that our MMMs exhibit comparable C3 separation performances with polycrystalline ZIF-8 membranes. The unexpectedly high C3 performance of the membranes was attributed to 1) unusually high ZIF-8 nanofiller loading in the selective skin layers (up to 50 wt%) and 2) improved molecular sieving properties of the ZIF-8 nanofillers (size less than 10 nm) resulting from the hindered linker flexibility upon confined filler formation.

12:00 pm (Room Palm AD, Tuesday)

Oral 66 - Effect of PPE decontamination processes on the filtration efficiency of porous polysulfone flat sheet membrane

Ebuka Ogbuoji (University of Kentucky)*; Isabel Escobar (University of Kentucky)

The ongoing Covid 19 pandemic has led to growing demands for face masks around the world for protection against the infectious Sars-Cov-2 virus. 3M, a Personal Protective Equipment (PPE) manufacturing company, reported a global production of two billion respirators in 2020. The most common material used in producing these commercially available face masks and respirators is polypropylene (a non-biodegradable thermoplastic) because of its chemical resistance, lightweight, and low cost. To reduce the waste caused by the excessive but essential use of face masks, studies have proposed implementing various decontamination techniques to ensure the safe reuse of facemasks. These techniques have been reported to reduce the mask filtration efficiency (FE) and compromise the mechanical properties of the mask material. This work involves the fabrication of a breathable high FE porous flat sheet polysulfone membrane, by nonsolvent induced phase separation (NIPS), resistant to common decontamination processes for face mask manufacture, which could potentially address the limitations of reusing nonwoven facemasks. Air permeation and pressure drop tests conducted across membrane samples showed lower resistance to airflow and pressure drop across membranes made from solutions containing PEG additive. The FE of the highest airflow membrane was observed to be 99.5% for particles greater than 300μm after testing using aerosolized NaCl particles at 35LPM. The breathable membrane, N95, and surgical masks were washed with detergent for 1 hour and dried to estimate the effect of a common and simple decontamination process. We observed an insignificant effect of washing with detergent on the fabricated membrane sample FE; however, there was a 20-25% reduction in N95 and surgical mask FE. The decrease in FE was due to depleted surface charges responsible for high FE in electret filter media, which functions by electrostatic interaction mechanism.

Oral 12 – BIOINSPIRED AND BIOMIMETIC MATERIALS

9:30 am (Room Palm BE, Tuesday)

Oral 67 - Artificial Water Channels-towards Biomimetic Membranes for Desalination

Mihail Barboiu* (Institut Europeen des Membranes)

Aquaporin (AQP) channels are efficiently water translocating proteins, rejecting ions. Inspired by this masterpiece of nature, Artificial Water Channels (AWCs) with controlled functional structures, can be potentially used to mimic the AQPs to a certain extent, offering flexible avenues toward biomimetic membranes for water purification. The objective of this paper is to trace back historical development and significant advancements of current reported AWCs. Meanwhile, we attempt to reveal important structural insights and supramolecular self-assembly principles governing the selective water transport mechanisms, toward innovative scale-up methods for industrial production of AWC-based biomimetic membranes for desalination.

1. M. Di Vincenzo, A. Tiraferri, V.-E. Musteata, S.Chisca, R.Sougrat, L.-B. Huang, S. P. Nunes, M. Barboiu,Nat. Nanotechnol. 2021, 16, 190-196.

2. M. Di Vincenzo, et al. Proc. Natl. Acad. Sci. USA, 2021, 118(37), e2022200118

3. L.-B. Huang, M. Di Vincenzo, M. Göktuğ Ahunbay, A. van der Lee, D. Cot, S. Cerneaux, G. Maurin, and M. Barboiu, J. Am. Chem. Soc. 2021, 143, 14386-14393

4. L.-B. Huang, M. Di Vincenzo; Y. Li, M. Barboiu, Chem. Eur. J., 2021, 27, 2224-2239

10:00 am (Room Palm BE, Tuesday)

Oral 68 - Quantification of salt and proton rejection in a Flourofoldamer based artificial water channel

Laxmicharan Samineni (University of Texas at Austin)*; Matthew Skiles (University of Texas at Austin); Woochul Song (University of Texas at Austin); Manish Kumar (University of Texas at Austin)

The global stress of clean water scarcity has led to immense interest in desalination membranes over the last few decades. The current state-of-art polymeric desalination membranes face challenges concerning their salt rejection and energy requirement. Despite extensive efforts, synthetic polymer membranes with performance better than the fully aromatic polyamide thin-film composite membranes have not been successfully demonstrated yet. Biomimetic and Bioinspired membranes comprised of high-density uniform pore structures mimicking natural water channels (aquaporins) have been proposed as a promising alternative to synthetic membranes. The key to this technology is fabricating sub-nanometer scale pore structures that can selectively transport only water and reject solute ions. Accurate single channel water and ion permeabilities quantification for different pore architectures is important to enable seamless comparison of novel channels and estimate their potential performance. In this work, we used stopped-flow florescence spectroscopy to resolve permeability of protons and chloride ions through novel flourofoldamer based artificial water channels (AWCs). A combination of HTPS and Lucigenin dye quenching experiments on vesicular systems and Fluorescence Correlation Spectroscopy to count the number of channels inserted per vesicle were used in this study to quantify the proton and chloride ion permeability (ions/channels/second). Our analysis shows that proton and chloride permeability of the fluorofoldamer is 0.83±0.52 and 0.52±0.18 ions/channel/s indicating the efficient solute rejection of the proposed AWCs. Along with the ability to reject solute ions, these channels showed high water permeability (1.4×10-10 H_2O/s) which is encouraging for application in biomimetic desalination membranes.

10:30 am (Room Palm BE, Tuesday)

Oral 69 - Fast water transport through biomimetic reverse osmosis membranes embedded with peptide-attached (pR)-pillar[5]arenes water channels

Yu Jie Lim (Singapore Membrane Technology Center, Nanyang Environment and Water Research Institute, Nanyang Technological University)*; Kunli Goh (Nanyang Technological University); Gwo Sung Lai (Nanyang Technological University); Chiann Yi Ng (Nanyang Technological University); Jaume Torres (Nanyang Technological University); Rong Wang (Nanyang Technological University)

This study examined the feasibility and performance of a nanochannel-based biomimetic membrane (NBM) for brackish reverse osmosis (RO) desalination. Two types of peptide-attached synthetic nanochannels, (pR)-pillar[5]arenes (pRPH) and (pS)-pillar[5]arenes (pSPH), were incorporated into liposomes. pSPH is a diastereomer of pRPH and was used as a negative control (i.e. mutant) to pRPH in this work. The nanochannel-containing liposomes (e.g. pRPH-liposomes) were then immobilized into the active layer of the RO membranes via in situ interfacial polymerization on the top of a polysulfone support membrane to form NBM-pRPH membranes. To maximize the potential and benefits of the NBM-pRPH membrane, the physical characteristics of the polyamide layer was further tuned using some additives and the eventual membrane was named as NBM-pRPH-A. The NBM-pRPH-A membrane exhibited a water permeability of 6.09 L m^{-2} h^{-1} bar^{-1} and 98.2% NaCl rejection under a 15.5

bar applied pressure using 2000 mg L^{-1} as feed solution. The 62% flux increment with respect to the pristine control is postulated to arise from a thinner, less cross-linked (more free volume) and more hydrophilic active layer as well as the possible supplementary transport pathways of the pRPH-liposomes. The performance of the NBMs under differential feed pressures and temperatures further exemplifies the water permeation property of the pRPH nanochannels. Accordingly, the NBM-pRPH-A gave a water permeability higher than commercial RO membranes tested in this work (DuPont BW30 and Hydranautics ESPA2) as well as other RO membranes reported in the literature. This study provides a tangible foundation for the development of NBMs for brackish RO desalination.

11:00 am (Room Palm BE, Tuesday)

Oral 70 - Preparation of capillary flow membranes from wood veneers with advanced electron and ion microscopy characterization

Nadezda Prochukhan (Trinity College Dublin)*; Lucy Fitzsimmons (Trinity College Dublin); Michael Morris (Trinity College Dublin)

Membrane technology has become an ever expanding and extremely relevant field as waste separation becomes critical in halting the environmental crisis. In particular, bioinspired and bioderived membrane materials offer inherent advantages of biodegradability and abundance and pose minimal environmental hazards.

We have looked at wood veneers from various species (pine, beech, oak and walnut) which upon chemical modification can show varied capillary flow rates. The structure of wood is intrinsically porous and membrane-like which we took advantage of. Bleaching and further deposition of chemical ligands such as biocompatible short polymer chains like polyethylene glycol modify the capillary were used as chemical modifiers. Delignification can increase the flow 30 to 70-fold depending on the species from minutes per cm to a few seconds per cm. Deposition of other chemical moieties can slow down the flow depending on hydrophilic quality of the ligand.

We also demonstrate how novel advanced microscopies such as low voltage scanning electron microscopy (SEM) and helium ion microscopy (HIM) with charge compensation can be used. Furthermore, focused ion beam (FIB) coupled with SEM was used to reconstruct a portion of wood to model the flow and pore structure through volume.

We expect that our findings will open up new avenues for research possibilities and novel applications. Furthermore, the use of biomaterials throughout the filtration process can instill a drive towards sustainable bioeconomy.

11:30 am (Room Palm BE, Tuesday)

Oral 71 - Skin like biomimetic membranes for breathable protective fabrics

Hyeonji Oh (University of Texas at Austin)*; Yu-Ming Tu (The University of Texas at Austin); Benny Freeman (University of Texas at Austin); Manish Kumar (University of Texas at Austin)

As human skin does not provide perfect protection against life-threatening biochemical agents, protective fabrics are necessary in warfare, pandemics, and terrorist attacks. However, first responders in conventional protective clothing suffer from the heat stress induced due to the lack of the breathability of the material. Therefore, we propose to utilize biomimetic membranes to reduce heat stress while maintaining the protection. In this work, we dehydrate and use the biomimetic membranes in a dry environment (in the air) for the first time. We show the promise of developing next generation breathable protective fabrics by utilizing 2D nanosheets of the beta-barrel biological channel, outer membrane protein F (OmpF). It shows an excellent water vapor transport rate of 241 g h^{-1} m^{-2}, overperforming the commercial breathable fabric, Pertexsheid. Moreover, it achieves high water resistance along with a high protective capacity against small molecules (MWCO of 640Da) and biological agents (7 log removal of MS2 bacteriophage).

12:00 pm (Room Palm BE, Tuesday)

Oral 72 - A TiO_2-Au Janus membrane has potential to enable self-pumping flow by light.

Yuhang Fang (Purdue University)*; Jefferey Moran (George Mason university); David M Warsinger (Purdue)

Light-driven micro-swimmers, such as TiO_2-Au Janus spheres could utilize the electrophoresis and propel themselves to swim in liquid. Inspired by the self-propelling micro-swimmer, this work aims to find an effective structure of TiO_2 - Au Janus membrane to achieve the self-pumping characteristics in membrane. To do this, a series of 2D-axisymmetric models were built to solve Poisson-Nernst-Planck-Stokes equations in COMSOL. Three designs (i.e. TiO_2-Au, Au-TiO_2-Au and TiO_2-Au-TiO_2) were compared to examine the self-generated flow rate. According to the results, a TiO_2-Au nanopore with length of 5um and pore radius of 1um could achieve a self-pumping flow rate of 27.27um/s, while there is no significant net flow but internal circulated flow in either Au-TiO_2-Au or TiO_2-Au-TiO_2 pore. Furthermore, given a constrained pore length, decreasing the pore radius or repeating the number of TiO_2-Au units would reduce the self-pumping flow rate. Even though the derived results are still elementary, this study has proved the applicability of building light-control self-pumping membrane. A more extensive study might shed new light on using the self-generated electric field to achieve particle trapping or species separation.

Oral 13 – MEMBRANES FOR ELECTROCHEMICAL APPLICATIONS I

9:30 am (Room Palm CF, Tuesday)

Oral 73 - Deconstructing Proton Transport Through Atomically Thin Monolayer CVD Graphene Membranes

Piran Kidambi (Vanderbilt University)*

Selective proton permeation through the atomically thin lattice of graphene and other 2D materials offers new possibilities for energy conversion/storage and novel separations. Practical applications necessitate scalable synthesis via approaches such as chemical vapor deposition (CVD) that inevitably introduce sub-nanometer defects, grain boundaries and wrinkles, and understanding their influence on H^+ transport and selectivity for large-area membranes is imperative but remains elusive. Using electrically-driven transport of protons and other ions through monolayer CVD graphene we systematically deconstruct selective proton transport. Mitigation of large defects, wrinkles and tears via interfacial polymerization results in selective proton permeation rates comparable to state-of-the-art proton exchange membranes e.g. Nafion. Our work provides a new framework to assess H^+ conductance and selectivity of large-area 2D membranes for practical applications.

10:00 am (Room Palm CF, Tuesday)

Oral 74 - Investigation of Porous PVDF Membranes Behavior during Metal-Assisted Chemical Imprinting

Aliaksandr Sharstniou (Arizona State University); Shouhong Fan (University of Colorado at Boulder); Emmanuel Dasinor (Arizona State University); Yifu Ding (Membrane Science, Engineering, and Technology (MAST) Center, Department of Mechanical Engineering, University of Colorado Boulder); Bruno Azaredo (Arizona State University)*

Current advancements in the field of nanophotonics require development of new manufacturing processes that will combine (i) sub-100 nm resolution, (ii) 3D geometrical control of the fabricated structures and (iii) high manufacturing throughput. Recently developed Metal-Assisted Chemical Imprinting (Mac-Imprint) is a contact-based wet etching process, which has the potential to meet the above requirements. In this process, noble metal-coated stamp is brought in contact with semiconductor substrate in the presence of etching solution. The stamp's pattern is transferred onto substrate due to the etching reaction, which is localized at the interface of noble metal/substrate contact. Large-area (i.e. wafer scale) format of the Mac-Imprint can be realized with the aid of fully porous stamps, which provide volume for the etching solution storage and sustain the diffusion of the reactants. The transition of Mac-Imprint to wafer scale, however, requires understanding of the behavior of such stamps under the etching conditions. Thus, this work investigates etching solution permeability, pore size and surface roughness of several types of porous PVDF membrane (commercially available and lab-made) as a function of noble metal film thickness, contact force and exposure to the etching solution.

10:30 am (Room Palm CF, Tuesday)

Oral 75 - Dialysate regeneration via photooxidation and FO membrane protection-loop to enable portable kidney dialysis

Guozheng Shao (Univ. of Wash.); Bruce Hinds (Univ. of Washington)*

Kidney disease effects 37 million people in the US with end stage renal failure requiring life-saving kidney dialysis that consumes 7% of the US annual medicade budget. In-center hemodialysis is the most prevalent (90%) but suffers from poor 5 year morbidity of 65% and severely reduced quality of life due to being tethered to a fixed dialyzer for 4 hours thrice weekly with associated non-physiological toxin build up. Ideal is to develop a small portable device that operates continuously, however this is prevented due to conventional dialysis session requiring 120 kg of purified water with added electrolytes. This can be reduced to about 1kg if the dialysate can be regenerated. We had previously demonstrated that a UV-TiO_2 based Photo-Oxidation Urea Removal (POUR) system [DOI: 10.1021/acsanm.9b00709] removed 14gr/day of urea to regenerate dialysate. However, careful analysis of spent dialysate species showed that the POUR unit would have to be 3 times larger to oxidize other organics such as glucose and amino acids. Hence a urea selective membrane would be required to increase the POUR unit efficiency. Lack of urea rejection with osmotic membranes (RO/FO) has been a long-standing problem in wastewater treatment, but this urea selectivity directly benefits dialysate regeneration. Using physiological mixtures of dialysate species, a compact RO unit showed rejections of 20% for urea, 80% for creatinine, 90% alanine, and greater than 98% for the 7 other most prevalent uremic species. Commercially available hollow fiber FO membranes (Aquaporin, 2.3m^2) with a 10mM concentration gradient had a urea flux of 30gr/day that was more than twice that required and corresponded to a urea mass transfer coefficient of 4x10^-7 m/s. The FO-POUR system was validated using human spent dialysate and met required 14gr/day urea removal in a recirculating dialysate. An analytical model has been developed to design multi-loop systems for a variety of clinical treatment modalities.

11:00 am (Room Palm CF, Tuesday)

Oral 76 - Electrochemical oxygen pumps for mobile medical oxygen

Ralph A Bauer (Global Research and Development Inc.)*

In this work we demonstrate a scalable wet-chemical synthesis for dense homogeneous layers of platinum, gadolinium doped ceria, and yttria-stabilized zirconia that are >1000□ thinner than the state of the art. Use of these YSZ layers as solid electrolyte results in >300 C lower operating temperatures and much increased performance. Such thin layers for electrodes and electrolytes are accomplished through the formation of stable aqueous dispersions of non-agglomerated, 1…2 and 2…3.5 nm spherical particles, that are stable for at least 2 months (Pt), 12 months (GDC) and 18 months YSZ. The particles are formed by sonochemical precipitation in an aqueous Y and Zr salt solution are packed in thin layers that transform into 100% phase-pure cubic YSZ at an unprecedented low temperature of 325°C. These layers can be used as electrolytes in supported thin film SOFC and electrochemical O_2 pump structures. State of the art SOFCs require operation at ≥600°C with power densities of ~3500 W/m2. Thin film SOFCs were fully functional at 325°C and generated 5200 W/m^2 at 550°C.

State of the art O_2 electrochemical pumps require operation at ≥750°C and a voltage of 10...15 V, yielding 0.5 L/min/m^2. For the first time operating scaled up tubular geometry electrochemical oxygen pumps operating in continuous, cross flow conditions have generated 11 L/min/m^2 at 600°C and a voltage of 1 V.

11:30 am (Room Palm CF, Tuesday)

Oral 77 - Liquid separations membranes made of functionalized poly-ether ether ketone blended with polysulfone

Abelline K Fionah (University of Kentucky)*; Laura Brady (University of Kentucky); Kate Mclarney (University of Kentucky); Isabel Escobar (University of Kentucky)

The global water crisis growth has led to a tremendous increase in membrane technology research. Membranes are favored over many other technologies for water treatment because, in principle, they require no chemical additives, and can be used isothermally at low temperatures. Membranes that can reject salts, produce good permeate flux and require minimal cleaning are highly demanded. However, most of the synthesized membranes on the market have associated problems, such as membrane fouling, as well as high cost of synthesis, operation, and maintenance. There is a need for a more effective and sustainable approach for optimizing membrane performance in application to water treatment. This research focuses on increasing the surface charge and hydrophilicity of polysulfone (PSf) membranes by incorporating a sulfonate-functionalized poly-ether-ether-ketone (S-PEEK) into a PSf/N-Methyl-2-pyrrolidone (PSf/NMP) membrane. The sulfonation of the polyether-ether ketone provides a net increase in negative charge on the surface of the membrane that enables charge repulsion to take place, increasing the rejection of ions from water bodies. In this project, the effect of applied potential on the performance of S-PEEK: PSf/NMP membrane was evaluated. Varying voltages of 1.5, 2.5 and 5V were applied to the membranes in the presence of various salts (sodium chloride, calcium chloride, and potassium chloride salts). It was found that both the permeability and the selectivity of the membranes were improved when voltage was applied. Characterization of the of the as synthesized membrane was carried out using SEM, contact angle measurements, FT-IR, zeta potential measurements, and cloud point analysis. The obtained results indicated that, incorporating S-PEEK into the PSf/NMP membrane, increased the hydrophilicity of the membrane, and under applied voltage, allowed it to function as electrodialysis capable of removing contaminants from water bodies by charge repulsion of ions.

12:00 pm (Room Palm CF, Tuesday)

Oral 78 - Lithium brine purification and electrochemical conversion via selective bipolar membrane electrodialysis

Aubrey Quigley (University of Texas at Austin); Kevin Reimund (University of Texas at Austin)*; Benjamin Pedretti (University of Texas at Austin); Kristofer Gleason (University of Texas at Austin); Nathaniel Lynd (The University of Texas at Austin); Benny Freeman (The University of Texas at Austin)

As the demand for lithium batteries increases, new domestic opportunities for lithium recovery from geothermal brines in the United States have attracted attention as a potential alternative to brine pond extraction processes widely practiced in South America, for example. However, geothermal brines are complex and contain contaminants such as magnesium, calcium, and sodium, with divalent ions being difficult to remove via ion sorption processes. Selective ion-exchange processes have been developed for upgrading the quality of such brines, but these methods typically produce a chloride salt product which is subsequently precipitated as Li_2CO_3, or electrochemically converted into the more-desirable LiOH in a separate process, such as bipolar membrane electrodialysis. Direct one- or few-step upgrading of Li-containing brine to LiOH is desired to avoid unnecessary processing of LiCl. Here, we report on the development of Li-selective cation-exchange membranes which enable the simultaneous concentration, electrochemical conversion, and removal of contaminants via selective bipolar electrodialysis. This process enables the upgrading/conversion of lithium to be practiced on less-processed brines with higher monovalent and divalent ion contamination. Selectivity for lithium over magnesium is observed at the cation-exchange membrane with a separation factor of 8:1.

Oral 14 – HIGH SALINITY STREAMS, BRINE MINIMIZATION AND ZLD

9:30 am (Room Abbey South, Tuesday)

Oral 79 - High Pressure Reverse Osmosis Membrane Modeling and Characterization

Jeffrey R McCutcheon (University of Connecticut)*; Yara Suleiman (University of Connecticut); Danh Nguyen (University of Connecticut); Mi Zhang (University of Connecticut); Ramesh Bhave (Ornl); Ramanan Sankaran (Oak Ridge National Laboratory); Luke Malencica (Oak Ridge National Laboratory); Kyle Cole (Michigan State University); Mayur Ostwal (University of Connecticut); Sina Shahbazmohamadi (University of Connecticut); Menachem Elimelech (Yale University); Ying Li (University of Connecticut); Marcus Day (National Renewable Energy Laboratory)

High pressure reverse osmosis (HPRO) is an alternative to energy intensive thermal desalination technologies for brine concentration. HPRO can potentially reduce the energy consumption of brine concentration by 90% over thermal desalination systems. Operating HPRO is challenging as state of the art thin film composite (TFC) reverse osmosis (RO) membranes do not perform well at excessively high pressures. These membranes can undergo changes when exposed to excessively high pressure. These changes can be modeled and observed in the different layers of the TFC membrane. We break down our work by layer

The polyamide selective layer – This layer is ultra-thin to enable fast water transport but highly crosslinked to restrict solute transport. Under excessive pressure, this layer may undergo microstructure changes. In this work, we valuate the microstructure changes and quantify impacts on permeance and selectivity using molecular dynamics simulations.

Porous support layer - The support layer is typically comprised of that is formed onto a nonwoven polyester scrim layer. Under high pressure, these layers will compress, leading to a reduction in support layer permeance. We use X-ray microscopy to quantify features of compressed membranes (porosity, tortuosity) and to measure the individual permeance of each part of the support layer.

Our findings indicate that polyamide layers do exhibit microstructure changes and a narrowing of the free volume element sizes when under extreme pressures. We also show substantial compression in an RO membrane support layer under higher pressures with more compaction occurring in the polysulfone layer than in the polyester layer. We conduct some tests in the presence of a spacer and find that compression is more severe where the permeate spacer presses against the membrane. Such findings provide critical insight into the future design of HPRO membranes, spacers, and modules.

10:00 am (Room Abbey South, Wednesday)

Oral 80 - Selecting Fouling and Scaling Mitigation Strategies for Membrane Distillation Applications with Water Reuse Reverse Osmosis Concentrate

Luke Presson (University of Arizona)*; Mukta Hardikar (The University of Arizona); Varinia V Felix (The University of Arizona); Andrea Achilli (The University of Arizona); Kerri Hickenbottom (U. Arizona)

Membrane distillation (MD) for the treatment of concentrated brines has been limited in part by membrane fouling and wetting. This study provides new insight into the identification of fouling and scaling mechanisms and pretreatment strategies for mitigating MD membrane fouling of water reuse reverse osmosis concentrate (ROC).

Long-term bench-scale direct contact MD experiments were performed with real ROC with pretreatment. Biological activated carbon (BAC) and water softening were selected as pretreatment strategies to isolate the effects of calcium and organic compounds individually and the feed solution was analyzed throughout the experiments. Organic compounds were analyzed by High-Performance Liquid Chromatography (HPLC), and inorganic compounds were analyzed by ion chromatography and ion coupled plasma chromatography. Membrane autopsies and a mass balance on the feed solution suggest that inorganic scalants were the only detectable constituents of membrane fouling despite the high organic content in the feed solution. Results from the HPLC suggest that the organic matter is entirely hydrophilic, indicating that hydrophobic-hydrophobic interactions between the organics and the membrane is not a mechanism for membrane fouling. Furthermore, the removal of organic compounds by BAC filtration resulted in minimal improvements, whereas water softening resulted in improved water flux and minimal membrane fouling throughout the experiments. Finally, the increasing rate of flux decline observed supports the Cassie-Baxter state theory of membrane fouling in which the presence of air pockets at the feed-membrane interface slows the adhesion of precipitates to the membrane. The analyses outlined in this study can assist in selecting the appropriate fouling and scaling mitigation strategies for a wide range of MD applications.

10:30 am (Room Abbey South, Wednesday)

Oral 81 - Highly Charged Ion-Exchange Membranes for Treatment of Brines via Electrodialysis

Carolina Espinoza (University of Michigan)*; José Carlos Díaz (University of Michigan); Kevin Kim (University of Michigan); Jovan Kamcev (University of Michigan)

Electrodialysis (ED) is a promising technology for brine concentration. Implementation of ED for this challenging application requires ion-exchange membranes (IEMs) that are highly permeable and highly selective when contacted by brines. Commercial IEMs for treating concentrated salt solutions exist, but further improvements in IEM throughput and selectivity are needed for ED to be competitive in this space. A better understanding of the interconnection between membrane structure and transport properties is required to guide IEM development for brine concentration. Commercial IEMs used for brine concentration have a low degree of swelling. However, there is a poor understanding of ion transport in such materials, making it difficult to rationally design IEMs with desired transport properties. To fill this gap, we have successfully synthesized a series of cross-linked IEMs with varying degrees of swelling and extremely high fixed charge concentrations. The ion transport properties of the IEMs were measured at salt concentrations representative of brines. The IEM counter-ion/co-ion selectivity was extracted using ion concentrations and diffusion coefficients in the membrane. Overall, there was little change in selectivity as IEM swelling decreased, but counter-ion throughput decreased about one order of magnitude over the IEM swelling range that we explored. The IEM selectivity was significantly affected by the external salt concentration. We saw a decrease of approximately one order of magnitude in IEM selectivity when the external salt concentration increased from 1 to 5 molal NaCl. The overall IEM selectivity was decoupled into ion sorption and diffusion selectivity to better understand the cause for the selectivity decrease. For all IEMs and salt solution concentrations, the ion sorption selectivity had a much greater contribution to the overall membrane selectivity than the diffusion selectivity. Strategies for improving the IEM selectivity will be discussed.

11:00 am (Room Abbey South, Wednesday)

Oral 82 - Characterizing low-salt-rejection RO membranes with the Kedem-Katchalsky model

Abdessamad BELGADA (Yale University)*; Yuhao Du (Yale University); Li Wang (Yale University); Menachem Elimelech (Yale University)

Low-salt-rejection reverse osmosis (LSRRO) is a newly developed membrane process for brine management. Recent studies demonstrated that LSRRO is capable of producing hypersaline brines with a moderate energy consumption compared to thermal brine concentrators. Understanding the salt-water separation mechanisms in LSRRO membranes is critical to optimize and further improve LSRRO technology. In this study, the Kedem-

Katchalsky (KK) model is employed to describe salt transport in LSRRO. In contrast to the solution-diffusion model that has been widely used for membrane desalination processes, the KK model considers both advective and diffusive salt transport. Additionally, the KK model allows the assessment of LSRRO membrane performance from experimental data through the determination of the phenomenological reflection coefficient and salt permeability. Specifically, by varying the exposure time of RO membranes to a chlorine solution, LSRRO membranes with different transport properties and salt rejections were obtained. The LSRRO membranes were then tested in filtration experiments using different feed salt concentrations (0.1 to 0.5 M NaCl) and various hydraulic pressures (14 to 33 bar). The phenomenological coefficients were extracted by fitting the KK model to the experimental data of salt rejection and water flux. Our results reveals that the reflection coefficient decreases and salt permeability increases as the exposure time of the membrane to the chlorine solution increases. Overall, our study demonstrates that the KK model can adequately describe salt transport in LSRRO membranes and thus can be used for the design of LSRRO systems.

11:30 am (Room Abbey South, Wednesday)

Oral 83 - Screening for scaling resistant desalination membranes

Sankaranarayanan Ayyakudi Ravichandran (University of Colorado Boulder); Saied Delagah (U.S. Department of Interior); John Pellegrino (University of Colorado)*

We have used both pressure-driven filtration and shaken-jar (non-filtration) testing to screen thin film composite (TFC) salt-rejecting membranes against surface adhesion of both calcium carbonate and calcium sulfate precipitating from supersaturated solutions containing other salts. We included homemade TFCs made with either a novel, mono-fluorinated trimesyl chloride (TMC) monomer or a non-fluorinated TMC polymerized with m-phenylene diamine. Pressure-driven filtration in crossflow mode caused supersaturation to be achieved over time and highly localized near the membranes' surface until a precipitous drop in permeation occurred. After which, the ability to recover the membranes' permanence by rinsing alone was also measured. The shaken-jar tests immediately exposed the membranes' surfaces to a supersaturated solution that gradually formed crystals and ultimately were suspended either in the bulk solution or strongly adhered to the membranes' surfaces. We measured surface scale deposition; crystal production; and overall Ca material balances to identify how precipitated crystals partition between the bulk and the surfaces of the TFC membranes with different surface chemistries. We determined that neither TFC membrane exhibited flux decline when exposed to solutions supersaturated with $CaCO_3$ despite micrographs indicating the presence of adhering scale on the membrane surface. In the case of $CaSO_4$ desupersaturating, both membranes exhibited flux decline and hydrodynamic cleaning with the supersaturated solution did not recover the flux. Similarly, both the membranes promoted the same desupersaturation in the shaken-jar measurements with equivalent, comparable levels of surface-adhered versus bulk suspended partitioning of the crystals.

12:00 pm (Room Abbey South, Wednesday)

Oral 84 - Optimizing pore size and charge of polyamide nanofiltration membrane for lithium enrichment from salt lake brine

Juan Zhai (Texas Tech University)*; Sunday P Adu (Texas Tech University); Yuexiao Shen (Texas Tech University)

Since demand for lithium is rapidly increasing globally, lithium extraction has been shifted to salt lake brines, which contain usually 100-400 ppm lithium. One of the biggest challenges for lithium extraction from such brine is separating Li^+ from Mg^{2+}, one of the representative coexisting divalent cations. This is because Li^{1+} and Mg^{2+} have similar hydrated radius and both are positively charged. Nanofiltration (NF) membranes are capable of separating Li^+ from Mg^{2+} based on synergetic size and Donnan exclusion mechanisms. In this work, we propose to tune pore size and surface charge in order to achieve a high separation factor of Li/Mg. We expect our membranes can achieve a Li/Mg separation factor of 20 while maintaining a moderate permeability of ~10 liter/($m^2 \cdot h \cdot bar$). We will also evaluate our membranes using synthetic brine and investigate the enrichment of lithium when multiple cations coexist.

Oral 15 – PROCESS INTENSIFICATION AND INTEGRATION

9:30 am (Room Abbey North, Tuesday)

Oral 85 - Assessment of a hybrid forward osmosis-freeze crystallization process for liquid mining of lithium: experimental study and numerical simulation

Afshin Amani (Université Laval)*; Georgios Kolliopoulos (Université Laval)

Forward osmosis (FO) is a concentration method that relies on the osmosis principle to selectively allow water to permeate through a semi-permeable membrane, while rejecting the solutes. This breakthrough technology could be considered as a sustainable liquid mining process for dilute solutions containing valuable metals, which due to their low concentration in solution cannot be recovered. Therefore, concentration by FO might be a promising direction towards the sustainable beneficiation of such streams. In this study, we report on the use of a hybrid forward osmosis -progressive freeze crystallization (FO-PFC) process for the liquid mining of lithium. Lithium chloride was considered as the feed solution (FS) in FO, at a concentration of circa. 3000 ppm, which is typical of brine lakes in North and South America and East Asia. Three types of chloride salts (NaCl, $CaCl_2$, and $MgCl_2$) were considered as the draw solution (DS) and the effect of the DS initial concentration on both the water and reverse solute flux was studied. PFC was in turn evaluated to regenerate the DS as an energy efficient process. The impact of temperature on the water recovery yield and purity was also investigated. Finally, the FO-PFC hybrid process was simulated using computational fluid dynamics with COMSOL Multiphysics software. Simulated results were in good agreement with our experimental data (<6% error), which enabled us to estimate parametric parameters accurately. Overall, our results indicate a strong potential for a hybrid FO-PFC process in liquid mining applications.

10:00 am (Room Abbey North, Tuesday)

Oral 86 – Thermal Storage Management Integrated with Solar Driven Membrane Distillation for Process Intensification

Mikah M Inkawhich (University of Arizona); Jeb C Shingler (University of Arizona); Spencer Ketchum (University of Arizona); Wei Pan (DWP Energy Solutions); Robert Norwood (University of Arizona); Andrea Achilli (The University of Arizona); Kerri Hickenbottom (U. Arizona)*

Management of concentrate streams in inland applications has an uncertain long-term environmental impact. This study investigates the efficiency of thermal energy management within an intensified air-gap membrane distillation (AGMD)-concentrated solar power (CSP)/photovoltaic (PV) collector. The demonstration-scale system has the potential to realize self-sustained zero-waste discharge of concentrate streams in inland and off-grid applications and can produce up to 180 kWh of thermal energy directly supplied to the AGMD system to produce up to 290 L/day. Experiments were performed on the hybrid system to evaluate thermal performance under various operating conditions including MD and CSP flow rates, thermal storage reservoir pre-heating, AGMD operating temperature, and AGMD vacuum pressure. Experimental results indicate that doubling the MD flowrate results in 116% increase in thermal energy utilization and nearly doubles the distillate production. Compared to the winter months, operating the hybrid system in summer months when direct normal irradiance (DNI) is at its peak results in a sixfold increase in average distillate production (30 L in winter and 180 L in summer). Furthermore, the relative specific thermal energy consumption decreases by 30% when allowing the thermal storage reservoir to preheat in the winter. To increase thermal storage capacity, energy utilization, and total distillate production the hybrid system was retrofitted to include an additional thermal storage tank with smaller volume and automated valves to operate in different configurations depending on environmental conditions. Compared to results obtained on the original hybrid system, the updated thermal storage configuration increased MD distillate production by approximately 28% in an initial winter operation comparison. Results from this study highlight important design considerations for integrating thermal desalination with solar energy resources in an operational environment.

10:30 am (Room Abbey North, Tuesday)

Oral 87 - Reverse Osmosis vs. Electrodialysis: Comparing the Cost of Brackish Water Desalination

Sohum Patel (Yale University)*; Boreum Lee (Yale University); Menachem Elimelech (Yale University)

Effectively combatting water scarcity requires harnessing unconventional, typically saline, water sources via desalination. As our reliance on desalinated water grows, the identification and implementation of the most effective technology — both in terms of cost and energy consumption — is critical towards ensuring sustainable water production. In a previous work, we focused on assessing the thermodynamic energy efficiency of desalinating brackish waters (salinity range of ~1 g L^{-1} to 10 g L^{-1}) with the two most prevalent technologies — reverse osmosis (RO) and electrodialysis (ED). Here, we extend on this technological comparison by including the economics of each process to provide a more holistic assessment. Specifically, we use rigorous technoeconomic models to perform a systematic and direct comparison of the levelized cost of water (LCOW) for RO and ED across a wide range of desalination conditions. By maintaining the same separation parameters (i.e., feed salinity, salt removal, water recovery) and productivity among each technology, we ensure that our comparison is valid and fair. We demonstrate that though ED has superior energy efficiency to RO in many low feed salinity applications, its LCOW in many of these cases is higher. Furthermore, we determine specific operating ranges in which each technology excels, accounting for potential enhancements in process performance and reduction in system costs. By performing a sensitivity analysis on each of the ED process parameters (i.e., membrane cost, membrane charge density, spacer thickness, membrane thickness, and power cost), we reveal the most promising avenues for reduction of the LCOW of ED. Our results indicate that cost reduction of ion-exchange membranes is currently the most impactful approach for making ED more competitive with RO for brackish water desalination.

11:00 am (Room Abbey North, Tuesday)

Oral 88 - Biocidal Effects In Microwave Induced Membrane Distillation

Indrani Gupta (New Jersey Institute of Technology)*; Sagar Roy (New Jersey Institute of Technology); Somenath Mitra (New Jersey Institute of Technology)

Pathogenic microorganism contamination in water is a threat to public health causing diseases such as dysentery, typhoid and cholera and such outbreaks have led to millions of deaths worldwide. Methods for treating pathogen-contaminated water include boiling, chlorination and ozonation, UV irradiation, and filtration. These methods can be expensive and can also produce disinfection byproducts that themselves are toxic. Recently, membrane processes which have low cost, smaller footprint, and provide effective treatment options for a variety of pollutants have become popular in water purification. In techniques such as nanofiltration and ultrafiltration, the efficiency of pathogen removal depends upon the size-exclusion property of the membrane and reverse osmosis (RO) membranes are prone to fouling.

Membrane distillation (MD) is a separation method where the driving force is the difference in vapor pressure between a hot feed side and a cold permeate side. While many pathogens get killed at higher temperatures, the challenge arises in the cases of those microorganisms that are stable at high temperatures namely thermophilic bacteria. It is well known that microwave irradiation kills pathogens. To enhance the efficiency of killing thermophilic bacteria, we implemented a microwave heating system in MD, and we refer to it as Microwave Induced MD (MIMD). It was observed that the biocidal performance as well as the flux was improved due to thermal and non-thermal factors of microwave irradiation. Non-thermal effects had a greater contribution towards higher cell destruction due to the interaction of the microwave with cellular matter disrupting secondary and tertiary structures, and towards improved water vapor flux (30 to 40%) due to localized superheating and enhanced hydrogen bonding break down of water molecules. Furthermore, MIMD required much lesser (20-25%) energy than conventional MD to carry out the experiments under the same conditions.

11:30 am (Room Abbey North, Tuesday)

Oral 89 - Automated and field-deployable evaluation of microfiltration/ultrafiltration fouling

Weiming Qi (Clemson University)*; David Ladner (Clemson)

The advantages of membrane filtration compared with other treatment methods are smaller footprints, high quality permeates, and full automation compatibility. However, membrane fouling limits filtration performance through complex interactions between the membrane surface and the foulant in the feed stream. Multiple efforts like frequent cleaning, membrane coating or grafting, and optimized filtration parameters have been tested to overcome membrane fouling. Most of this research requires a reliable membrane filtration system that can be operated in various modes; however, it is hard to find a detailed description of filtration systems used in literature since most papers only roughly described the system in the methods section and many systems are not fully automated.

This presentation aims to provide details about building a lab-scale filtration system with automated control. This system allows small-scale, high-throughput investigations of membrane processes that more adequately mimic full-scale operations than traditional laboratory experiments. The automated filtration system is coupled with a membrane module with associated hardware and a data acquisition/control program. The program created in LabVIEW collects data from different sensors and processes them to maintain the operating target. Frequent backwashing and cleaning in place (CIP) are automated, triggered with different criteria for optimizing the energy use and decreasing chemical and permeate consumption. The automation also allows remote, un-manned evaluation of water quality fluctuations in industrial processes, as demonstrated by a deployment we accomplished during the COVID pandemic. Other functions like step-flux functions, flux/TMP targetting, and data processing fulfill the requirement of different types of membrane experiments.

12:00 pm (Room Abbey North, Tuesday)

Oral 90 - Scale-up of membrane distillation for treating produced waters from oil and gas industry.

Ritesh Pawar (University of Pittsburgh)*; Zhewei Zhang (University of Pittsburgh); Radisav Vidic (Pitt)

The recent increase in hydraulic fracturing for the extraction of oil and gas from unconventional reservoirs led to an increase in the amount of high salinity produced water generated in this process. Because of the highly distributed nature and varying characteristics of produced waters across the shale plays, there is a unique opportunity of treating these waters by using modular Membrane Distillation (MD) technology. The TDS, specific inorganic ions and organic contents vary greatly depending on the region and lifetime of the wells. In some cases, the pretreatment required for the removal of inorganics and organics plays a vital role in determining the feasibility of the overall treatment success. The study focuses on pilot-scale demonstrations of Air Gap MD technology using the produced water from different shale plays, including Permian Basin (TX), Bakken formation (ND), and Marcellus shale play (PA). The pilot-scale demonstration in Permian Basin was performed using fairly simple pretreatment of raw produced water (i.e., filtration, chemical oxidation, and precipitation) to achieve 120 h of steady-state operation without any operational problems. Greater than 99.7 % salt rejection was obtained with 0.8 GPM feed flow rate and at 50 % water recovery. Similarly, the pilot-scale system was successfully operated with 90 and 180 g/L produced water from the Bakken region to achieve water recovery of 60 and 30 %, respectively. These tests have demonstrated that it is possible to recover high-quality permeate from a complex and high salinity produced water, thereby effectively reducing the volume of produced water that would need disposal and reducing the environmental footprint of the unconventional oil and gas industry. Similar tests are planned with produced water from the Marcellus region. Additionally, cost assessment of treating produced water with different salinities and recoveries along with various pretreatments options will be carried out.

Oral 16 – POLYMERIC AND MIXED-MATERIALS – GAS SEPARATION II

2:00 pm (Room Palm AD, Tuesday)

Oral 91 - Reduction of physical aging in copper nanoparticle pillared CMSMs derived from a polyimide precursor incorporating the bis(phenyl)fluorene-based cardo moiety and uniformly dispersed metal organic polyhedra-18 (MOP-18)

Masoumeh Tajik Asl (The University of Texas at Dallas)*; John Ferraris (University of Texas at Dallas); Kenneth Balkus (Univerisity of Texas at Dallas); Inga Musselman (University of Texas at Dallas)

Carbon molecular sieve membranes (CMSM) were prepared by carbonizing precursor polyimides incorporating the bis(phenyl)fluorene-based cardo moiety. The bulky cardo diamine in precursor prevents tight packing of the structure and restricts the ring rotation which enhances the gas permeability. By introducing the cardo moiety in the precursor polymer structure, the permeability has been increased from ~6000 in Matrimid to ~14700 in BTDA-BAF.

CMSM suffer from physical aging due to the densification of membranes over time, resulting in pore collapse and a decrease the permeability of membranes. We report the results of pillaring with copper nanoparticles derived from the thermal decomposition of a uniformly dispersed soluble metal organic polyhedra-18 (MOP-18), which prevent collapsing of the large unconnected graphene domains in carbonized BTDA-BAF.

2:30 pm (Room Palm AD, Tuesday)

Oral 92 - Novel graphene oxide-based membrane structure for a highly effective breathable barrier for toxic vapors and chemical warfare agents

Yufeng Song (New Jersey Institute of Technology); Cheng Peng (New Jersey Institute of Technology); Zafar Iqbal (New Jersey Institute of Technology); Kamalesh K Sirkar (New Jersey Institute of Technology)*; Gregory Peterson (4CBR Filtration Branch, R&T Directorate DEVCOM Chemical Biological Center, U.S. Army Futures Command)

Garments protective against chemical warfare agents (CWAs) or accidentally-released toxic chemicals must protect from toxic gases/vapors by blocking their transport for substantial time and allow moisture transport for breathability. A number of approaches have been studied. A thick mixed matrix membrane having a high loading of metal organic framework HKUST-1 in an elastomeric triblock copolymer completely blocked 2-chloroethyl ethyl sulfide (CEES) (a simulant for sulfur mustard) for quite some time (i.e., high value of permeation lag time) but the moisture vapor transmission rate (MVTR) was low~192 g/m^2-day (minimum needed, 1500-2000 g/m^2-day). Using a highly breathable carbon nanotube (CNT) based membrane blocking biological agents, an alternative approach grew on the top surface of the CNTs a layer of polymer chains which reversibly collapsed in contact with a nerve agent simulant diethyl chlorophosphate (DCP) temporarily shutting CNT pore mouths and dramatically reducing the permeation rate of DCP. However, DCP permeation was restricted to $\approx$1% of the chemical challenge. Complete blockage for an extended period during use is absent. Our initial studies using a cross-linked graphene oxide (GO) laminate achieved 98.3% rejection of CEES on a steady basis after a short permeation lag time1. Our newly developed GO based laminate structure however shows remarkable results namely, complete blockage of ammonia, Mustard (HD), Soman (GD), and a Sarin simulant, dimethyl methyl phosphonate, for 2750 min, 1075 min, 176 min, and 7 days respectively while moisture transmission rate was substantial. We will describe this structure and its performance.

1Peng, C., Iqbal, Z., Sirkar, K. K., Peterson, G. W., Graphene oxide-based membrane as protective barrier against toxic vapors and gases, ACS Appl. Mater. Interfaces. 2020, 12, 11094−11103.

3:00 pm (Room Palm AD, Tuesday)

Oral 93 - Peculiar effect of low loading of metal-organic polyhedra on CO_2/N_2 separation properties of cross-linked polyethers

Taliehsadat Alebrahim (The State University of New York at Buffalo)*; Liang Huang (The State University of New York at Buffalo); Heshali Welgama (The State University of New York at Buffalo); Shiwang Cheng (Michigan State University); Durga Acharya (CSIRO); Cara Doherty (CSIRO); Anita Hill (CSIRO); Clayton Rumsey (The State University of New York at Buffalo); Martin Trebbin (The State University of New York at Buffalo); Tim Cook (University at Buffalo); Haiqing Lin (The State University of New York at Buffalo)

Mixed matrix materials (MMMs) containing porous fillers merge the excellent processability of polymers with the well-controlled pores and strong size-sieving ability of the fillers. Gas permeability of MMMs is often described using the Maxwell model, and high loadings are often required to maximize the benefits of the porous fillers. Herein, we demonstrate that MMMs comprising low loading of metal-organic polyhedra (MOP) and cross-linked polyethers exhibit unexpectedly high gas permeability compared with the Maxwell model. MOPs are discrete nanocages with high porosity and excellent compatibility with polymers. To prepare MMMs, the MOPs were first dispersed in prepolymer solutions containing macromonomers of polyether before photopolymerization. Several MOPs are studied, including Cu-MOP, Fe-MOP, and Zr-MOP, and ether macromonomers studied include poly(ethylene glycol) diacrylate (PEGDA), poly(ethylene glycol) methyl ether acrylate (PEGMEA), and poly(1,3 dioxolane). As the Cu-MOP content increases from 0 to 3 wt.% in PEGDA-co-PEGMEA, CO_2 permeability increases from 510 Barrer to 1000 Barrer while retaining the high CO_2/N_2 selectivity at 35 °C; a further increase of loading to 5 wt.% decreases CO_2 permeability to 730 Barrer without influencing CO_2/N_2 selectivity. By contrast, at the Cu-MOP loading of 10 wt.% or above, the gas permeability can be satisfactorily described by the Maxwell model. Similar behavior have been observed for MMMs containing low content of Fe-MOP, and Zr-MOP. According to SAXS measurements, there is a strong low-q upturn at q-values down to 10−3 Å−1 showing that MOP nanofillers form percolation networks even at low loadings around 2 and 3 wt.% which can be linked to the high permeability at low loadings. These MMMs were thoroughly characterized using FTIR, WAXD, SAXS, DSC, and SEM. The MMMs with low loadings and much-improved permeability can be easily integrated into current manufacturing processes of thin film composite membranes.

3:30 pm (Room Palm AD, Tuesday)

Oral 94 - Cross-linked poly(ionic liquid)-ionic liquid composite membranes for CO_2/light gas separations: mixed-gas separation performance and long-term stability under high-temperature and -pressure conditions

Chamaal Karunaweera (University of Colorado Boulder)*; Douglas Gin (Department of Chemistry, University of Colorado - Boulder); Richard Noble (Deparment of Chemical and Biological Engineering, University of Colorado - Boulder)

CO_2/light gas (CH_4 and N_2) separation performance, physical aging and plasticization of cross-linked poly(ionic liquid)-ionic liquid (PIL-IL) composite membranes were evaluated at elevated temperatures (up to 50 □C) and feed pressures (up to 40 atm). PIL-IL composite membranes made with IL-based multifunctional cross-linkers exhibited higher CO_2 permeability and CO_2/light gas selectivity compared to the membranes made with a conventional, non-ionic cross-linker, divinylbenzene (DVB). We also show that the gas separation performance of PIL-IL membranes can be further tuned by varying the wt % of cross-linker, PIL:IL ratio, and both simultaneously. For example, a cross-linked PIL-IL composite membrane prepared using 70 wt % 1-vinyl-3-methylimidazolium bistriflimide as the IL monomer, 30 wt % 1-ethyl-3-methylimidazolium bistriflimide as the free IL, and 8 wt % 1,3,5-tris(1'-methylene-3'-vinylimidazolium bistriflimide)benzene as the IL cross-linker exhibited a ca. 50% increase in CO_2 permeability and a comparable CO_2/CH_4 selectivity compared to an analogous PIL-IL-zeolite MMM cross-linked with DVB. In addition, the same PIL-IL membrane exhibited 90% retention of its initial CO_2 permeability after testing at 50 °C and 40 atm pressure for 42 weeks (7100 h). In comparison, an analogous PIL-IL composite membrane cross-linked with DVB shattered after testing for 5 h at the same testing conditions. These cross-linked PIL-IL composite membranes can also be formed into thin film composite (TFC) membranes. The PIL-IL TFC membranes exhibited enhanced permeance compared to free-standing PIL-IL membranes.

4:00 pm (Room Palm AD, Tuesday)

Oral 95 - Directing Flexibility of Metal-Organic Framework toward Crystallographically Derived Molecular Sieving

Ke Zhang (Aramco Americas)*

Metal Organic Frameworks (MOFs) revitalize interests for polymer-inorganic mixed-matrix membranes with their structural versatility and tunable pore systems for potential membrane performance improvements. Contrary to relative rigid zeolites, MOFs usually exhibit extraordinary flexibility that results in a lack of sharp molecular sieving by their nominal pore structures. From the perspective of kinetic separation, it is very challenging to restrain MOF flexibility while delivering molecular cut-off windows delineated by the nominal aperture sizes of MOF materials. In this talk, discussions will be focused on some findings about how polymer matrix may be utilized to restrain MOF flexibility and deliver kinetic separations determined by MOF crystallographic aperture size. ZIF-8 is a well-studied ZIF with nominal aperture size (~3.4 Å) but is demonstrated to be most effective in kinetically separating propene and propane with an effective aperture size of 4.1 Å due to its extraordinary framework flexibility. Here, new results will demonstrate that, by restraining framework flexibility, ZIF-8 MMMs can exhibit superior performance for small gas separations that are far beyond their respective upper bound limits (with molecular sieving in the range of 3.4-3.6 Å), while the signature ZIF-8 propene/propane selectivity is no longer observed. Together with a strong plasticization resistance, this membrane platform offers opportunities for restraining MOF flexibility and enhancing membrane transport properties for relevant industrial gas separations.

4:30 pm (Room Palm AD, Tuesday)

Oral 96 - Ethylene and Ethane Transport Properties of Hydrogen-stable Ag^+-based Facilitated Transport Membranes

Matthew N Davenport (The University of Texas at Austin)*; Caitlin Bentley (The University of Texas at Austin); Joan Brennecke (The University of Texas at Austin); Benny Freeman (The University of Texas at Austin)

Previous studies on Ag^+-containing facilitated transport membranes report high olefin-paraffin selectivity (e.g., ethylene over ethane) but poor stability of the Ag^+ carrier, especially in the presence of reducing gases, such as H_2, H_2S, and C_2H_2 which are often present in industrial process streams. Solid polymer electrolytes consisting of crosslinked poly(ethylene glycol) diacrylate and up to 70 wt% silver bis(trifluoromethylsulfonyl)imide ($AgTf_2N$) salt are synthesized through a facile and scalable UV-crosslinking process. Following over 10 weeks of pure H_2 permeation at 4 bar and 35°C, the membrane retains over 90% of its initial pure gas ethylene-ethane selectivity. X-ray photoelectron spectroscopy does not detect a change in the oxidation state of the dissolved Ag^+ ions. At the highest $AgTf_2N$ concentration (70 wt%), pure-gas ethylene-ethane selectivity and permeability are 21 and 4.0 Barrer, respectively, yielding performance surpassing the polymeric gas separation membrane upper bound. Dual-mode ethylene sorption is observed and modeled using a chemical equilibrium model, and gas diffusivity is calculated according to the solution-diffusion model. Preliminary results concerning mixed-gas transport properties in addition to membrane stability in the presence of other reducing agents, namely H_2S and C_2H_2, are also included.

Oral 17 – DOWNSTREAM BIOPROCESSING

2:00 pm (Room Palm BE, Tuesday)

Oral 97 - Membrane adsorber purification of pDNA

Herb Lutz (Milliporesigma)*; Thomas Elich (Milliporesigma); Tyler Cheung (Milliporesigma)

Plasmid DNA (pDNA) is an essential raw material for making gene therapy viral vectors and mRNA therapeutics. Chromatographic purification poses unique challenges compared to protein purification. Because the size of pDNA is large, access to pores of traditional adsorptive resin beads is hindered, resulting in low binding capacity. Chromatography membranes offer a more open pore structure, providing an opportunity for improved pDNA purification capacity and productivity. This talk will consider capture purification of a commercial pDNA feedstock using a membrane adsorber currently on the market.

The capture step employs anion exchange adsorbers (AEX) in bind-elute mode to concentrate pDNA, remove RNA, and reduce endotoxin. Loading at high salt enables the RNA to pass through in the flow-through fraction while binding pDNA. The Natrix® Q AEX membrane adsorber features large 0.4 µm pores which bind pDNA with higher capacity and at much shorter residence times (on the order of seconds) compared to a commercial resin (on the order of minutes). A wash step is used to improve the purity of the bound pDNA. The high capacity and ability to rapidly cycle adsorber modules debottlenecks processes to satisfy demand, and shrinks the footprint required for reduced consumables and buffers for a more sustainable process.

Because membrane adsorbers are offered in discreet holder-less sizes, linear scale up of modules for large batches requires plumbing in a parallel assembly. Module-to-module permeability variation can potentially lead to a flow distribution issue different from resin column packing. This talk will show how this effect can be readily overcome. As nucleic acids gain prominence as a new therapeutic modality, membrane adsorbers are available to meet the purification challenge.

2:30 am (Room Palm BE, Tuesday)

Oral 98 - Predicting the transport of soft droplets in porous media from measurable emergent properties

Guillaume Lostec (University of Colorado-Boulder); Franck Vernerey (University of Colorado-Boulder); John Pellegrino (University of Colorado)*

Understanding the transport of soft particles (droplets, colloids, vesicles, cells) in porous media is relevant to drug delivery, medical diagnostics, water treatment as well as food and cosmetic industries. Toward this objective we previously illustrated a multiscale modeling approach, connecting the mechanics of particles deforming through individual pores to the structure of the porous medium. The experimental validation of this model and its predictive potential are yet limited by the lack of a calibration procedure. Indeed, the model microstructure is highly idealized and the parameters describing its geometry cannot be directly measured on a real membrane. To address this issue, we present a method to find the value of the model's parameters such that the idealized geometry can be considered equivalent, in terms of soft particles transport, to a real porous medium. We first review the characteristics of the model with an emphasis on the input parameters and their influence on particles transport. Then, we introduce our calibration process based on matching the value of a set of measurable emergent properties in the model with a real membrane. We present a phase diagram predicting the transport of soft particles through a specific membrane and how it can be obtained with the model calibration.

3:00 pm (Room Palm BE, Tuesday)

Oral 99 - Purification Of Viral Vectors With Novel Membrane Adsorbers

Cristiana Boi (Università di Bologna); Jinxin Fan (North Carolina State University)*; Ruben Carbonell (North Carolina State University)

Given the recent growth in gene and cell therapies and to the great potential of the Adeno-associated virus (AAV) as a delivery vector of therapeutic genes for multiple diseases, we developed a purification process using strong anion exchange membrane chromatography. In this study, we present the purification of AAV2 from a Sf9 lysate that was used as a starting material. The membranes consisted of glycidyl methacrylate (GMA) grafted polybutylene terephthalate (PBT) nonwoven fabrics that were further functionalized by introduction of a quaternary ammonium (Q) group. Membranes were prepared using two different methods of grafting: UV grafting (UVG) and heat induced grafting (HIT). The membranes were first characterized for their capacity to bind AAV2 at different residence times (RTs) using a pure AAV2 full-capsid standard. When loading the pure viral particles at 1.0 min RT, the binding capacity of Q-PBT(UVG) reached 9.6E+13 viral particles (vp)/mL, a capacity higher than those measured for Q-PBT(HIG) and a commercial Sartobind Q unit (0.8 mL membrane bed volume).

Purification of AAV2 Sf9 lysate using the Q membrane confirmed the high binding capacity of the of Q-PBT(UVG) nonwoven membranes with this complex supernatant. The AAV2 recovered from our Q-PBT membranes (both UVG and HIG) exhibited a higher purity than the Sartobind Q membrane. For improving purity, a preliminary purification with iminodiacetic acid cation exchange nonwoven (IDA-PBT) was used, and the eluate treated with the Q membrane. The AAV2 product captured from the Sf9 lysate using this process had a similar purity to the pure standard AAV2 product and the overall two-step membrane chromatography process achieved a high recovery of infectivity of 76.0 %. This work provides new insights into viral particle bind and elute behavior on the nonwoven-based ion exchange membrane adsorbers and demonstrates that these membrane adsorbers can be used for efficient purification of AAV vectors.

3:30 pm (Room Palm BE, Tuesday)

Oral 100 - The Effects of Low Flow and Flow Disruption on Virus Filtration

Wenbo Xu (University of Arkansas); Ranil Wickramasinghe (University of Arkansas); Xianghong Qian (University of Arkansas)*

Virus filtration is routinely conducted under the normal filtration mode at constant pressure. Continuous bioprocessing has significant potential advantages with improved productivity, better product quality and a smaller foot-print. Commercial virus filters typically have a defined product throughput and an upper limit on virus loading. Viral clearance may become inadequate if these limits are exceeded. Continuous biomanufacturing will require current practice of virus filtration to adapt. Virus filtration will most likely be run under low, constant flux conditions and possibly with a processing hold. However, there is little knowledge on the performance of virus filters in low-pressure constant flux mode with regard to capacity and virus retention. Here the mechanisms for virus migration through the pores of the commercially available virus filters were investigated at constant flux. Confocal imaging studies were also performed to visualize the locations where the virus particles were captured at different constant fluxes. Results indicate that back-diffusion at low-pressure and low flux condition could lead to the permeation of virus particles through the filter. On the other hand, at high filtration flux, the hydrodynamic force dominates leading to the entrapment of the virus particles inside the filter.

4:00 pm (Room Palm BE, Tuesday)

Oral 101 - High Purification of Binary Protein Mixtures having Close Molecular Weights by Ultrafiltration

Lixin Feng (New Jersey Institute of Technology); Yufeng Song (New Jersey Institute of Technology)*; Sagnik Basuray (New Jersey Institute of Technology); Kamalesh K Sirkar (New Jersey Institute of Technology); Solomon Isu (University of Arkansas); Ranil Wickramasinghe (University of Arkansas)

Mixtures of proteins/biopharmaceuticals having close molecular weight ratios (MWRs) are generally separated by chromatography. Conventional ultrafiltration (UF) can separate binary protein mixtures if the MWR of the two proteins is at least ~7–10. For lower MWRs, novel countercurrent cascades of 3-5 stages can yield relatively pure proteins. To separate similarly sized proteins in one device, High Performance Tangential Flow Filtration (HPTFF) technique was developed. HPTFF achieves high selectivity but does not yield almost pure protein in the permeate. The Internally Staged Ultrafiltration (ISUF) technique was proposed to yield almost pure protein in the permeate. The ISUF technique uses a stack of 3 UF membranes without any gaskets/spacers in-between. Permeate from membrane 1 with selective skin facing the feed mixture becomes the feed to the next membrane 2 located below and so on. We have investigated separations of a number of binary protein mixtures using the ISUF configuration in a stirred cell: (1) hemoglobin (MW 64,677 Da) and BSA (MW 66,430 Da) having a MWR of 1.03; (2) α-lactalbumin (14,400, MW) and β-lactoglobulin (18,400, MW; present naturally as a dimer, 36,800) mixture having a MWR of 2.55 and encountered in whey protein concentrate separation. In each case, we have investigated the effect of pH, ionic strength and feed pressure using as required 100kDa or 30kDa MWCO Ultracel® and Omega® membranes having a particular charge. We have investigated the effect of diavolumes on the purity of the retained protein as well as other performance indicators. We have also investigated the purification of a protein from a virus using a stacked set of UF membranes. Virus filtration for validation of virus clearance is difficult when the size of the mAb approaches that of small virus particles.

4:30 pm (Room Palm BE, Tuesday)

Oral 102 - Understanding the mechanisms that govern B. diminuta microfiltration to improve the sterile filtration of therapeutic viruses

Evan Wright (McMaster University)*; Alexandra Jucan (McMaster University); David Latulippe (McMaster University)

Sterile filtration processes based on the use of 0.2 µm rated microfiltration membranes are commonly used in biopharmaceutical downstream processing to satisfy regulatory requirements that the final product be free of contaminating bioburden. The pore size rating of those membranes is designated based on complete retention of the bacteria B. diminuta (according to the ASTM F838 protocol). It is well known that the use of 0.2 µm rated microfiltration membranes delivers low yields (<50%) of many viral vectors that are used in cell and gene therapies and this is a large cost driver in virus manufacturing. Thus, there is strong motivation to developing microfiltration processes that maximize the yield of viral vectors while also satisfying the criteria for B. diminuta retention.

In this study, we compared the performance of nine commercial microfiltration membranes with different chemistries (PES, PVDF, cellulose acetate) and pore size ratings (0.2, 0.45, 0.8 µm) in filtration experiments with B. diminuta in saline lactose broth (as per ASTM F838 protocol). We found that some of the 0.45 µm rated membranes completely retained B. diminuta at low operating pressures (~ 5 psi); most interesting was that the performance of a PES 0.45 µm membrane from one supplier was radically different than of a PES 0.45 µm membrane from another supplier. Zeta potential measurements on the nine membranes showed very little difference in their charge properties. Thus, it seems that there is a complex relationship between the transmission of B. diminuta and physico-chemical properties of microfiltration membranes. We also compared the performance of 0.2 and 0.45 µm rated membranes in terms of the yield during filtration of a rhabdovirus-based vector. The results from this work provide important insights into bacteria transmission through microfiltration membranes and guidance on the judicious selection of sterile filtration conditions to obtain high yields of viral vector products.

Oral 18 – MEMBRANE FOR ELECTROCHEMICAL APPLICATIONS II

2:00 pm (Room Palm CF, Tuesday)

Oral 103 - Separation of Rare Earth Elements from Simulated Geothermal Water using Novel Electrodialysis Metathesis Process

Lucy M Camacho (Texas A&M University-Kingsville, Department of Environmental Engineering)*; Mohammad U. Shafiq (Texas A&M University-Kingsville)

Rare earth elements (REE) demand has considerably increased in the last decade due to their uses in many current technologies such as touchscreen, microchips, magnets, and catalysts. The demand will continue to increase while supply becomes critically short. Therefore, identifying new sources and methods to extract REE becomes paramount. Separation of REE from simulated geothermal waters was conducted using a novel electrodialysis metathesis (EDM) process, which expands on typical electrodialysis by including a metathesis reaction to increase the selectivity of the ions and solubility of the concentrate streams while helping to reduce membrane scaling. In this study, the effect of pH on the solubility of REE was studied prior to conducting EDM experiments. A correlation between solution pH and the amount of precipitate recovered was found, with higher pH values producing larger amounts of the precipitate. The EDM separation of REE from simulated geothermal waters was analyzed using a 23 Factorial Design. The design analyzed the effect of voltage, REE concentration, and membrane type on the separation of Lanthanum, Cerium, and Europium. MINITAB® was used to analyze the factorial design using analysis of covariance (ANCOVA) within the general linear model (GLM). The research identified voltage as the only parameter having a significant effect on the separation of REE. These results indicated that the separation and concentration of REE from simulated geothermal water can be efficiently accomplished using the EDM process.

2:30 pm (Room Palm CF, Tuesday)

Oral 104 - Electrocatalytic Upcycling of Nitrate Wastewater to Ammonia Fertilizer via Electrified Membrane

Jianan Gao (New Jersey Institute of Technology)*; Yifan Li (Qingdao University of Technology); Bo Jiang (Qingdao University of Technology); Taha Marhaba (New Jersey Institute of Technology); Wen Zhang (New Jersey Institute of Technology)

Recovery and repurposing of wastewater nitrogen such as nitrate (NO_3^-) and nitrite (NO_2^-) receive increasing attention as the current approaches of biological nitrogen removal or conversion into nitrogen gas (N_2) renders high energy footprints. This study reported the first electrified membrane upcycling of NO_3^- into NH_3 that further converted into ammonia sulfate (NH_4SO_4), a liquid fertilizer readily for use. Paired electrolysis was used to couple simultaneous cathodic and anodic electrochemical half-reactions to enable nitrate reduction and eliminate the external acid/base consumption. Under a partial current density of 63.8±4.4 $mA \cdot cm^{-2}$ on a cathodic membrane made of a mixed-valence copper oxide and a polytetrafluoroethylene (PTFE) hydrophobic substrate, a recovery rate and energy consumption of 3100±91 g-$(NH_4)_2SO_4 \cdot m^{-2} \cdot d^{-1}$ and 21.8±3.8 $kWh \cdot kg^{-1}$-$(NH_4)_2SO_4$ were achieved with the synthetic feed solution (150 mM NO_3^-) flowing into the cathode chamber and the produced NH_3 migrating across the cathodic membrane into a trap and anode chambers filled with the flowing 0.5 M Na_2SO_4 solution. 99.9% NO_3^- was removed in the feed after 5 h operation with a NH_3 recovery rate of 99.5%. This electrified membrane process was demonstrated to achieve comparable performances of synergistic nitrate decontamination and nutrient recovery using real nitrate wastewater with durable catalytic activity and stability.

3:00 pm (Room Palm CF, Tuesday)

Oral 105 - Co-transport of lithium, sodium, and potassium ions in polysulfone-based CEMs

Jung Min Kim (University of Virginia)*; Sean M Bannon (University of Virginia); Geoffrey M Geise (University of Virginia)

Lithium (Li^+) is among the most valuable ions for energy storage devices (e.g., lithium-ion batteries). Li^+ mining is one of the most common approaches for global lithium production, but it has environmental concerns such as excessive use of freshwater (e.g., ~2 kg of water is needed to extract 1 g of lithium). Lithium-rich geothermal brines (e.g., that found at the Salton Sea) are recognized as more eco-friendly sources of lithium and could represent a low carbon footprint Li^+ source. However, the composition of these brines makes Li^+ recovery difficult. Electromembrane processes could contribute significantly as a component of a broader separation scheme to isolate Li^+ from complex solutions containing other ions. A major challenge facing this approach is to separate Li^+ from other monovalent ions, and Na^+ and K^+ are two prime examples. Sulfonated polysulfone (i.e., sulfonated biphenol-based polysulfone (sBPS)) is a promising material for cation exchange membranes (CEMs) with favorable thermal and mechanical properties. Poly(ethylene glycol) (PEG), a linear polymer containing ether oxygen, produces a strong dipole moment toward cations that hydrate strongly, suggesting favorable interactions with Li^+. Engineering Li^+ selective PEGylated sBPS CEMs by combining both of these materials, to leverage their favorable properties, could address challenges facing electromembrane-based monovalent ion separations. Ion transport properties were measured using both single salt measurements (i.e., LiCl, NaCl, or KCl) and measurements where mixtures of the cations were exposed to the membrane. The results inform both the Li^+ selectivity of the modified sBPS material with respect to Na^+ and K^+ and phenomena that occur when the membrane is challenged with mixtures of ions.

3:30 pm (Room Palm CF, Tuesday)

Oral 106 - Zwitterions And Their Influence Block Polymer Morphology

Bradley Grim (Arizona State University)*; Matthew Green (Arizona State University); Frederick Beyer (Army Research Laboratory).

Ion-containing block polymers (BPs) have seen significant interest in several membrane applications ranging from energy storage and conversion to gas and liquid purification. The morphology of BPs plays a substantial role in the mechanical and transport properties of BP membranes and has thus been a growing area of inquiry. The introduction of ionic species into BPs has been shown to greatly shift and skew the phase diagram relative to the neutral case. Ion translational entropy promotes block mixing which degrades performance and defeats the purpose of using a BP. Covalently binding both ions to the backbone helps minimize ion translational entropy and hypothetically improve segregation strength, however, our findings indicate this to be less trivial. In this work, we use an entirely ether-based platform tethered to either a common pendant charge with a mobile counterion or a zwitterion where both ions are covalently bound to study the influence of counterion entropy on BP thermodynamics. To our knowledge, this is the first time that backbone-tethered zwitterions have been used to study BP thermodynamics in the bulk and one of very few studies on bulk BPs involving zwitterions at all. Using dielectric relaxation spectroscopy, small angle X-ray scattering, atomic force microscopy, and rheology, we seek to understand how counterion entropy impacts BP thermodynamics by coupling microphase morphology and segregation strength to ion dynamics and interactions.

4:00 pm (Room Palm CF, Tuesday)

Oral 107 - Recent Progress In High Capacity, Stable, Lithium-Sulfur Batteries

Matthew R Hill (CSIRO and Monash University)*

One of the great global scientific challenges is the availability of energy at low cost, to all. Whilst the ability to generate power from renewable sources, and from a distributed grid, progressed remarkably over the last 15 years, it can only be fully exploited if usage can be decoupled from generation. Electrochemical energy storage technologies such as batteries and supercapacitors are of course the primary mechanism to break this link. Developments in the performance of new storage media, whilst maintaining an economical aspect, are crucial for making the use of self-generated renewable energy viable in many more instances.

One of the most tantalizing battery chemistries in this regard is Lithium-Sulfur (Li-S). With a gravimetric storage potential around 10 times higher than current Lithium ion technology, Li-S batteries are also manufactured from cheap, safe and stable materials, amenable to their widespread deployment. However, Li-S batteries have inherent challenges of stability with cycling, accessing their full storage potential, and obtaining charge-discharge at the requisite rates for the applications envisaged. Many of these inherent challenges are due to slow, or non-selective, transport within the cell, and porous materials offer the potential to address this behavior.1

In this presentation, recent breakthroughs in addressing these challenges inherent to Li-S will be presented. We have been able to develop Li-S cells with some of the best performances yet reported, and under conditions practical to their wide deployment. This is due to new forms of transport being achieved, and the mechanisms behind how porous materials within Li-S cells achieved this, discussed.

4:30 pm (Room Palm CF, Tuesday)

Oral 108 - Diffusion coefficients of condensed counterions in ion-exchange membranes: application of screening length scaling relationship in concentrated electrolytes

Yuxuan Huang (Columbia University)*; Hanqing Fan (Columbia University); Ngai Yin Yip (Columbia University)

Advancing fundamental understanding of membrane transport phenomena in ion-exchange membranes (IEMs) can enhance the performance of the membranes in environmental and energy applications. The recently proposed counterion condensation theory offers quantitative predictions of ion partitioning and co-ion diffusivity in IEMs. However, the simulation of counterion diffusivity is impeded by the lack of a condensed phase diffusivity model. This study presents a theoretical model for predicting condensed counterion diffusivity in IEMs, in which the screening length scaling relationship in concentrated electrolytes is employed to account for the screening length within the membrane matrix. The diffusivities of different condensed counterions in commercial IEMs were determined from experimentally characterized membrane ionic conductivity and the Donnan-Manning model. The results imply that the condensed counterions, which were assumed to be immobile in the traditional counterion condensation framework, are actually mobile and migrate under an externally applied electric field. The condensed counterion diffusivity model yields reasonably good predictions for monovalent ions with no adjustable parameters. However, larger deviations were observed for multivalent counterions, possibly attributed to the incomplete description of the screening length for ion-ion interactions. The study advances the understanding of fundamental IEM transport mechanisms and can facilitate the rational development of membranes with desirable ion transport selectivity.

Oral 19 – MEMBRANE CHARACTERIZATION

2:00 pm (Room Abbey South, Tuesday)

Oral 109 - Multi-lab study on the pure-gas permeation of commercial polysulfone (PSf) membranes: measurement standards and best practices

Katherine Mizrahi Rodriguez (MIT)*; Wan-Ni Wu (MIT); Zachary Smith (MIT)

Gas-separation membranes are a critical industrial component for a low-carbon and energy-efficient future. As a result, many researchers have been testing membrane materials over the past several decades. Unfortunately, almost all membrane-based testing systems are home-built, and there are no widely accepted material standards or testing protocols in the literature, making it challenging to accurately compare experimental results. In this multi-lab study, ten independent laboratories collected high-pressure pure-gas permeation data for H_2, O_2, CH_4, and N_2 in commercial polysulfone (PSf) films. Equipment information, testing procedures, and permeation data from all labs were collected to provide (1) accepted H_2, O_2, CH_4, and N_2 permeability values at 35°C in PSf as a reference standard, (2) statistical analysis of lab-to-lab uncertainties in evaluating permeability, and (3) a list of best practices for sample preparation, equipment set-up, and permeation testing using constant-volume variable-pressure apparatuses. Results summarized in this work provide a reference standard and recommended testing protocols for pure-gas testing of membrane materials.

2:30 pm (Room Abbey South, Tuesday)

Oral 110 - A Diafiltration Apparatus for High-Throughput Analysis of Membrane Transport Properties

Jonathan Ouimet (University of Notre Dame); Xinhong Liu (University of Notre Dame); Zachary Muetzel (University of Notre Dame); David Brown (University of Notre Dame); Elvis Eugene (University of Notre Dame); Tylar Popps (University of Notre Dame); Alexander Dowling (University of Notre Dame); Bill Phillip (Notre Dame)*

Creating systems and techniques capable of reducing the time, energy, and resources needed to characterize the transport properties of membranes can help to increase the rate of material and process development. In this study, guided by the tools of data science, a diafiltration apparatus is developed to rapidly characterizing membrane performance over a broad range of feed solution compositions. The apparatus doses a fixed-concentration diafiltrate solution into a stirred cell to achieve a predetermined change in the retentate concentration. Here, using an 80 mM potassium chloride (KCl) diafiltrate solution, it was shown that membrane performance, within a 5 mM to 80 mM KCl phase space, could be probed ten times more quickly with one diafiltration experiment (4 h) than with an experimental campaign using traditional filtration processes (47 h). Additionally, the synergy between data analytics and instrumentation led to the incorporation of an inline conductivity probe that monitored the real-time retentate concentration. This additional information provided key insights to distinguish between the mechanisms that govern membrane separations (e.g., discriminating between adsorption or rejection-based separations) and allowed for the membrane transport coefficients to be determined accurately. Ultimately, the ability of this device to characterize functional membranes rapidly will help to address knowledge gaps related to the interfacial processes that govern solute–solute selectivity and the performance of membranes in complex multi-component feed streams.

3:00 pm (Room Abbey South, Tuesday)

Oral 111 - New Insights in Ion Transport in Polyamide Membranes Using Impedance Spectroscopy

Viatcheslav Freger (Technion - IIT)*; Mikhail Stolov (Technion - Israel Institute of Technology)

Thin-film polyamide composites dominate today's desalination, yet they still pose numerous questions regarding structure and transport of small molecules and ions. The complex morphology lacking homogeneity down to nanoscale and poorly defined geometry, chemistry and intrinsic (bulk) properties of the polymer makes it difficult to relate them to water and solute permeabilities in way that would allow consistent comparison with models.

This situation gives an advantage to directly looking at transport properties, as a way to gain new insights. Although they may still be difficult to convert to intrinsic characteristics, the observed dependence on variables, such as ion type, solution composition and pH, and temperature, may serve as fingerprints, elucidating, validating, or ruling out different transport mechanisms. Our group has used electrochemical impedance spectroscopy (EIS) to measure membrane conductance, a transport characteristic, complementary to conventional pressure- or concentration-driven permeation measurements. In addition, being an in situ technique, EIS may be used to monitor variations of ion transport properties upon exposure to different solutes that impair or modify polyamide performance, such chlorine or molecular plugs.

In this presentation, I will summarize our recent results and insights gained using EIS, which include

- The unexpected effect of membrane charge, controlled by pH-dependent dissociation of polyamide's charged groups
- Specific interaction with different ions, with particular emphasis on mono- and di-valent cations and protons
- Temperature effects and activation energies of permeation
- The use of EIS for understanding the effects of molecular plugs, small molecules that can potentially improve polyamide selectivity towards certain solutes, e.g., boric acid

These findings are discussed in relation to transport mechanisms, including ion solvation (dielectric exclusion) and association, whose importance we highlighted recently.

3:30 pm (Room Abbey South, Tuesday)

Oral 112 - Interferometry for precisely measuring ultralow flow rates from permeable materials

Cody Ritt (Yale University)*

Novel observations in nanofluidic systems, such as ultrafast water transport, have directed recent research efforts toward developing new materials for water purification. Characterizing the transport properties of these materials is often restricted to exceptionally small areas and low pressures to avoid damage or defect formation. Consequently, observable flow rates are often limited to a few nL/h, even for materials with intrinsically high permeability. In this work, we apply laser interferometry to indirectly measure extremely low flow rates from permeable materials. We first demonstrate the principles of interferometry for measuring the rate of a moving mirror. We then validate the technique for nanoscale liquid level displacements by measuring the rate of water evaporation from a cylindrical vessel before (0.40 nm/s) and after (undetectable) sealing the vessel. When applying interferometry to membrane characterization, we find good agreement with water permeance measured in a conventional pressure-driven cell (mean absolute error, MAE = 8.54%). This proof of concept, and the perspectives given herein, could help expedite investigatory science on the fluid flow properties of novel nanofluidic devices and materials critical to innovation at the water–energy nexus.

4:00 pm (Room Abbey South, Tuesday)

Oral 113 - The Influence of Amine Structure and Water on the Mechanism of CO_2 Facilitated Transport: Operando Characterization and Kinetic Modeling

Casey O'Brien (University of Notre Dame)*

Reactive amines in polymeric facilitated transport membranes facilitate CO_2 transport across the membrane to give rapid and selective CO_2 separation from mixed gas streams. However, the fundamental mechanisms by which the amine carriers facilitate CO_2 transport are not well understood. For example, it is not clear whether the amine group catalyzes the transformation of CO_2 into bicarbonate species, which are mobile within the polymer matrix, or if CO_2 hops from one amine group to another as a carbamate species, which is chemically attached to the polymer chain. Furthermore, water vapor as well as the structure of the amine group can influence CO_2 transport across the membrane. How exactly these factors influence the CO_2 transport mechanism is not well understood. In this talk, I will describe a new operando transmission FTIR spectroscopy permeation cell that we have designed and fabricated to detect the formation of carbamate, bicarbonate, and other species in polymeric FTMs and correlate formation of specific intermediates to CO_2 permeation rates measured simultaneously on the same sample. I will also describe a new kinetic model that we have derived to quantify the relative contributions of facilitated transport and solution-diffusion to the overall rate of CO_2 transport across the membrane. The utility of this new operando spectroscopy tool and kinetic modeling approach will be demonstrated by investigating the influence of amine structure (primary, secondary, and tertiary amines) and water vapor on the kinetics and mechanisms of CO_2 facilitated transport across polyvinylamine-based membrane. I will show that the CO_2 transport mechanism is sensitive to both the amine structure and water, which influences the rate and selectivity of CO_2 permeation. The mechanistic insights obtained from this new operando methodology with kinetic modeling will enable a more rational design of FTMs than conventional membrane testing with ex-situ characterization approaches.

4:30 pm (Room Abbey South, Tuesday)

Oral 114 - Inadequacy of current approaches for characterizing membrane transport properties at high salinities

Yuanzhe Liang (Stanford University)*; Alexander Dudchenko (SLAC National Accelerator Laboratory); Meagan Mauter (Stanford Univerisity)

Cost optimal design of osmotic membrane processes requires an accurate estimate of membrane transport parameters across their full operational range. However, standard approaches for estimating these parameters rely on empirical methods, the accuracy of which remains unquantified as a function of temperature, salinity, and measurement error. Herein, we present a systematic accuracy analysis of previously developed methods for estimation of membrane transport properties in reverse osmosis, high-pressure reverse osmosis, forward osmosis, pressure retarded reverse osmosis, and osmotically assisted reverse osmosis. We use a Monte Carlo approach to sample the full range of feasible membrane water permeabilities, salt permeabilities, structural parameters, and operating conditions for these processes. These material and process parameters are then incorporated into a physical transport model for each process. Our analysis shows that the statistical uncertainty of current empirical methods for estimating membrane parameters increases by 5 times from low-salinity to high-salinity conditions. The result of this work demonstrates that empirical methods are inadequate for precisely estimating membrane transport properties at high salinity and highlight a critical need for the development of statistically validated higher accuracy methods.

Oral 20 – MOLECULAR AND PROCESS MODELING

2:00 pm (Room Abbey North, Tuesday)

Oral 115 - The impact of ion-ion correlated motion on salt transport in solvated ion exchange membranes

Nico Marioni (The University of Texas at Austin)*; Zidan Zhang (The University of Texas at Austin); Harnoor Sachar (The University of Texas at Austin); Everett S Zofchak (The University of Texas at Austin); Sanket Kadulkar (The University of Texas at Austin); Tyler J Duncan (The University of Texas at Austin); Benny Freeman (The University of Texas at Austin); Venkat Ganesan (The University of Texas at Austin)

Ion exchange membranes are widely used in water purification and energy generation applications to selectively and efficiently regulate salt transport, yet the influence of non-ideal ion-ion correlated motion on salt transport in these systems is often neglected and remains poorly understood. In this work, we study the impact of ion-ion correlations on salt transport in hydrated polystyrene sulfonate membranes and aqueous salt solutions via atomistic molecular dynamics simulations. Our results suggest that in both systems, cation-anion motion is positively correlated and increases significantly with doped salt concentration due to increased ion association arising from favorable electrostatic interactions. Similarly, the motion of distinct anions is found to be positively correlated and increase with salt concentration in the membrane system, but distinct cation motion in membrane and both distinct cation and distinct anion motion in aqueous solution is negligibly correlated. We attribute these observations to the imbalance between the concentration of free (mobile) cations and anions and the retarding effect of the fixed charge groups on cations in the membrane system.

2:30 pm (Room Abbey North, Tuesday)

Oral 116 - Elucidating CO_2/Hydrocarbons Solubility Selectivity in Aliphatic Polycarbonates via DFT Calculation and Experimentation

Thien N Tran (University at Buffalo, The State University of New York)*; Yuqing Fu (University of California, Riverside); De-en Jiang (University of California, Riverside); Haiqing Lin (The State University of New York at Buffalo)

Polyethylene oxide (PEO)-based materials are of great interest for CO_2/gas separation due to their favorable interactions towards CO_2 with quadrupole moment. There is also a strong interest in identifying new functional groups beyond ether oxygens to further improve CO_2/gas separation performance. In this work, we investigate the potential of the polar carbonate groups for this application using an integrated DFT calculation and experimentation, as exemplified by two aliphatic polycarbonates, poly(ethylene carbonate) (PEC) and poly(propylene carbonate) (PPC). The binding energy between gas molecules and polymer analog was calculated via DFT to elucidate the interactions, and the results were compared with gas solubility measured experimentally. CO_2 exhibits more negative binding energy than hydrocarbons, such as CH_4, C_2H_6, and C_3H_8, confirming the favorable interactions between the carbonate groups and CO_2. Interestingly, CO_2 solubility in PPC and PEC is 1.8 and 1.5 cm3(STP)/cm3/atm, respectively, which can be ascribed to the higher free volume in PPC than PEC and is consistent with the binding energy, -19 and -16 kJ/mol for PPC and PEC analog, respectively. Both PEC and PPC show CO_2/gas solubility selectivity substantially greater than PEO. For example, PEC exhibits CO_2/C_2H_6 solubility selectivity of 24 at 35 oC, much higher than PEO (2.8). Furthermore, PEC and PPC show CO_2/CH_4 permeability selectivity of $\approx$53 at 35 oC, which is one of the highest reported for rubbery polymers. PEC also exhibits CO_2/N_2 selectivity of 74, one of the best in the literature. The DFT calculation can be useful in designing next-generation membrane materials for important gas separations, such as carbon capture and natural gas processing.

3:00 pm (Room Abbey North, Tuesday)

Oral 117 - Non-equilibrium statistical mechanics approach for extended solution-diffusion model

Albert Kim (U. of Hawai'i)*

Reverse osmosis (RO) is a pressure-driven membrane separation process to separate salt ions and pure water in the feed solution (e.g., seawater). Transport of water and solutes across the RO membrane has been well explained using the solution-diffusion model that characterizes a membrane using two parameters: water and solute permeabilities of A and B, respectively. The water and solute fluxes are proportional to the net hydraulic pressure (i.e., the applied pressure subtracted by the osmotic pressure) and the solute concentration gradient across the membrane. Recent studies indicated that the solute flux is proportional to the difference between squared concentrations of the upstream and downstream. This phenomenon is explained by combining mass transport and charge neutrality. The current study investigates the solute transport phenomena proportional to the squared concentrations using the non-equilibrium statistical physics to extend the conventional solution-diffusion model.

3:30 pm (Room Abbey North, Tuesday)

Oral 118 - Mechanisms of Ion and Water Transport in PEGDA Membranes

Everett S Zofchak (The University of Texas at Austin)*; Zidan Zhang (The University of Texas at Austin); Harnoor Sachar (The University of Texas at Austin); Nico Marioni (The University of Texas at Austin); Sanket Kadulkar (The University of Texas at Austin); Tyler J Duncan (The University of Texas at Austin); Benny Freeman (The University of Texas at Austin); Venkat Ganesan (The University of Texas at Austin)

Selective separation of monovalent cations from complex mixtures is an industrially relevant procedure necessary for the recovery of many commodity materials, such as lithium from salt brines. Unfortunately, most conventional membranes lack selectivity between monovalent ions, rendering their use in such applications infeasible. One approach to endowing membranes with ion-ion selectivity is to incorporate ion-polymer interactions into materials to bias the selective partitioning and or diffusivity of one species over another. However, little is known about the impact of such interactions on the mechanisms of ion and water transport. In this work, we use a combination of experiments (ion permeation, conductivity, PFG-NMR) and molecular dynamics simulations to assess the influence of ion-dipole interactions on cation (alkali), anion (halide), and water transport in a model membrane material, poly(ethylene glycol) diacrylate. Experimentally, the order of salt and cation diffusion coefficients for LiCl, NaCl, and KCl are seen to deviate from the aqueous solution order. Molecular simulations attribute this deviation to cation-EO coordination in the membrane. Both the fraction of bound cations and the average lifetime of binding is observed to increase with decreasing hydration free energy of cations (moving down the alkali series), leading to different diffusivity trends in the membrane compared to solution. Additionally, we observe that the self-diffusivity of water increases with salt concentration for NaCl and KCl but decreases for LiCL, which is distinctly different from the monotonic decrease in water self-diffusivity observed in aqueous solutions. Simulations correlate this increase in water diffusivity for NaCl and KCl to a decrease in the average size of PEO chains and thus an increase in free volume. Our results reveal critical molecular-level insights into the mechanisms underlying salt and water transport in hydrated PEO-based membranes.

4:00 pm (Room Abbey North, Tuesday)

Oral 119 - Decoding the molecular-scale determinants of antifouling at polymer membrane selective layers

Dennis C Robinson Brown (The University of California Santa Barbara)*; Joshua Moon (California Nanosystems Institute, The University of California Santa Barbara); Thomas Webber (The University of California Santa Barbara); Lynn Katz (The University of Texas at Austin); Rachel Segalman (The University of California Santa Barbara); Scott Shell (The University of California Santa Barbara); Songi Han (The University of California Santa Barbara)

Current water purification membrane technologies cannot readily treat produced water (PW) feed streams from oil and gas operations which contain high concentrations of numerous organic and inorganic foulants. At present, improving resistance to fouling relies on macroscale heuristics such as manufacturing smooth, charge neutral, and hydrophilic membrane selective layers. Hydrophilic coatings such as polyethylene glycol (PEG)-based hydrogels dramatically increase resistance to apolar organic foulants. This anti-fouling property is hypothesized to originate from the formation of a bound water layer at the membrane surface that resists adsorption of hydrophobic molecules; however, the molecular basis for hydrophilic anti-fouling is not well understood. In this work, we investigate the molecular-scale dynamics, structure, and solvation thermodynamics at the interface between water and PEG in bulk-solution and at model membrane surfaces using molecular dynamics (MD) simulations. In PEG-water mixtures ranging from 0 to 50 weight percent PEG, we observe quantitative agreement between water self-diffusivities measured via Pulsed Field Gradient Nuclear Magnetic Resonance (PFG-NMR) spectroscopy and atomistic MD simulations. Further, we reveal a striking correlation (R^2~0.95) between the equilibrium self-diffusivity of water and the enhancement of water's tetrahedral structure with PEG concentration. Finally, we quantify the solute-surface affinity of several common electrically neutral, PW constituents with varying hydropathy at a model PEG hydrogel surface. We find that the solute-surface affinity of these foulants resembles that of ideal hydrophilic surfaces. These results enable a new analytical framework to understand the molecular-level determinants of water properties and solute-polymer interactions at membrane selective layers and provide a path towards the rational design of anti-fouling water purification membranes.

4:30 pm (Room Abbey North, Tuesday)

Oral 120 - Towards a universal framework for evaluating mass transport in pressure, concentration, and temperature driven membrane-based desalination systems

Kian P Lopez (University of Colorado Boulder)*; Ruoyu Wang (vanderbilt university); Elizabeth A Hjelvik (University of Colorado Boulder); Shihong Lin (Vanderbilt University); Anthony Straub (University of Colorado Boulder)

Desalination technologies using salt-rejecting membranes are a highly efficient tool to augment existing water supplies via desalination. Such technologies rely on (1) a membrane that facilitates fast water transport but rejects solutes and (2) a thermodynamic driving force to effect water permeation. In recent years, numerous studies have worked to advance a variety of membrane processes with different membrane types (liquid- and air-filled) and driving forces (pressure, concentration, and temperature). These processes encompass a wide range of systems, from the technologically mature reverse osmosis process to emerging membrane distillation systems. Despite all these processes being used for water desalination, direct quantitative comparison of different technologies has been difficult, often showing contradictory results for water productivity and other key performance metrics. In this work, we critically assess different membrane-based desalination technologies and provide a universal framework for comparing various driving forces and membrane types. To accomplish this, we first introduce a simple equation to quantify the thermodynamic driving forces available from pressure, concentration, and temperature gradients across salt-rejecting membranes. We then examine the resistances experienced by water molecules as they traverse membranes with differing transport mechanisms and material properties, drawing a correlation between liquid and vapor permeability that allows for direct comparison. Next, we consider how temperature and concentration polarization impact desalination performance. Finally, we conduct simulations and literature analysis to quantitatively and qualitatively compare desalination processes, identifying scenarios where each process has advantages. Overall, this work demonstrates how mass transport between the different types of membranes can be directly related and applications where each desalination process can be most efficiently implemented.

Abstracts – Oral Presentations, Wednesday

Presenting authors are indicated by an asterisk*. See the most up-to-date version of the program book with full abstracts at **www.membranes.org/nams~2022/**

Oral 21 – AWARD SESSION

9:30 am (Room Palm AD, Wednesday)

Oral 121 - Diglycolamide membrane adsorbers to separate lanthanides and actinides for use in radiopharmaceuticals

Christine E Duval (Case Western Reserve University)*; Maura Sepesy (Case Western Reserve University); Alec Johnson (Case Western Reserve University); Tuli Banik (Case Western Reserve University); Megan Sibley (Case Western Reserve University); Joelle V Scott (Case Western Reserve University); Alexa Ford (Case Western Reserve University); Timothy Yen (Case Western Reserve University)

Actinium-225 is a promising radionuclide for delivering localized cancer treatments that can "seek and destroy" metastasized cancers. To date, all actinium-225 used in domestic clinical trials is purified from legacy nuclear waste at Oak Ridge National Laboratory. This supply is so limited that only 1 in 5,500 people who qualify can receive treatment. Current purification methods rely on 'extractive' resin-packed columns which are operated at low flow rates (<1 mL/min) due to the slow diffusion of ions to binding sites. Slow purification (6-8 hours per column) creates a bottleneck in the Ac-225 production process and limits the US supply of this in-demand isotope. Innovations in purification materials are needed to decrease purification time and advance more cancer therapies to market. This contribution describes our efforts to synthesize a new class of acid-stable, radiation-stable membrane adsorbers for use in medical isotope purification.

In this work we 1) synthesize new, acid-stable membranes, 2) synthesize diglycolamide (DGA) ligands with reactive end groups and 3) covalently tether the DGA ligands to the membrane surface. Membranes are synthesized by electrospinning poly(styrene-co-chloromethyl styrene) fiber mats. Membranes are characterized for their surface area, fiber diameter and porosity. DGA derivatives are synthesized with reactive end groups (amines or alkenes) and structures are confirmed with high-resolution mass spectrometry, elemental analysis, and nuclear magnetic resonance spectroscopy. DGA derivatives are covalently bound to the membranes by coupling reactions. In our initial experiments, lanthanum (a surrogate for actinium) is used to determine the binding capacity and binding kinetics of the DGA-functionalized membranes. Lanthanum concentrations are determined by inductively coupled plasma optical emission spectroscopy. The results of this research are paving the way for faster separations in radiochemistry.

10:00 am (Room Palm AD, Wednesday)

Oral 122 - Putting bubbles to work: Next-generation water treatment systems using air-trapping membranes

Anthony Straub (University of Colorado Boulder)*

Current reverse osmosis (RO) membranes are vulnerable to degradation from oxidants commonly used in water treatment, poorly remove low molecular weight neutral solutes, and are constrained by the permeability-selectivity trade-off. Despite many attempts to develop improved RO membranes, their material design has remained nearly unchanged for four decades. Here, we present the development of a novel class of RO membranes that utilize vapor transport through sub-100 nm pore size hydrophobic films. We show that these membranes offer near complete removal of contaminants that pass through conventional RO membranes including boron, urea, and disinfection byproducts. We also find that reducing thickness allows the membranes to increase permeability without compromising salt rejection, potentially enabling such membranes to circumvent the permeability-selectivity trade-off that limits conventional polymeric RO membranes. The developed membranes also resist oxidation from chlorine and ozone, and we demonstrate new concepts for water treatment enabled by oxidation-resistant ultra-selective membranes. Scale-up of the desalination method using large-area polymeric membranes is demonstrated. Overall, this work presents a novel class of RO membranes that shows advantages in selectivity, productivity, and fouling resistance.

10:30 am (Room Palm AD, Wednesday)

Oral 123 - Elucidating the role of micropore generating backbone motifs and amine functionality on sorption energetics, mixed-gas permeation, and H_2S transport

Katherine Mizrahi Rodriguez (MIT)*; Pablo Dean (MIT); Naksha Roy (MIT); Sheng Guo (MIT); Timothy Swager (Massachusetts Institute of Technology); Zachary Smith (MIT)

While significant headway has been made in the synthesis of microporous polymers for gas separations, little is known about their structure–property relationships in industrially relevant conditions involving condensable gases such as H_2S. Our recent work on the mixed-gas transport in an amine-functionalized microporous PIM (PIM-NH_2) has demonstrated surprising benefits of plasticization resistance and competitive sorption for increasing selectivity in binary CO_2/CH4 mixtures. Here, we test the generalizability of this approach through the synthesis of an amine-functionalized polymer based on a poly(aryl ether) (PAE) backbone. The amine-functionalized PAE (PAE-NH_2) exhibits a 3-fold increase in CO_2/CH_4 and CO_2/N_2 mixed-gas permselectivities compared to the pure-gas case. In 20/20/60 $H_2S/CO_2/CH_4$ mixtures tested at a total pressure of 8 atm, amine-functionalized derivatives sustain increases in CO_2/CH_4 selectivities compared to nitrile-functionalized analogues. Mixed-gas performance trends were analyzed through ternary mixed-gas sorption modeling, dual-mode sorption analysis, and investigation of sorption energetics for N_2, CH_4, CO_2, and H_2S. CO_2 sorption for PAE-NH_2 was significantly higher than that of PAE-CN. The strength of the gas–polymer interactions were quantified through calculation of isosteric heats of sorption for all gases in PIM-1, PIM-NH_2, PAE-CN, and PAE-NH_2. Amine-functionalized films had highly exothermic interactions with CO_2 with heats of sorption around –44 kJ mol–1. PIM-NH_2, PAE-CN, and PAE-NH_2 demonstrated excellent plasticization resistance in 50/50 CO_2/CH_4 mixed-gas tests up to 26 atm, and increased stability to pure-gas H_2S at higher pressures compared to PIM-1. Our results indicate the development of highly CO_2-selective polymers for CO_2 capture and natural gas purification. The structure-property-performance relationships derive in this work provide fundamental and practical design handles for the future development of membrane solutions.

11:00 am (Room Palm AD, Wednesday)

Oral 124 - Selective ion-ion separations and desalination using adsorptive ion-capture electrodialysis membranes

Adam A Uliana (University of California, Berkeley)*; Ngoc Bui (The University of Oklahoma); Jovan Kamcev (University of Michigan); Mercedes Taylor (University of Maryland); Jeffrey Urban (Lawrence Berkeley National Laboratory); Jeffrey Long (University of California, Berkeley)

The removal of trace yet toxic constituents (e.g., heavy metals, boron) from water presents an especially pressing technological challenge. Established materials and water treatment processes seldom meet the performance standards needed to efficiently isolate these pollutants for proper disposal or re-use. To address this issue, this work presents the development of a membrane-based water treatment strategy we call "ion-capture electrodialysis." In this approach, water is desalinated through an electrodialysis mechanism while targeted ions are simultaneously captured by selective adsorbents embedded within ion-exchange membranes. We apply multifunctional cation-exchange membranes dispersed with Hg^{2+}-selective, thiol-functionalized porous aromatic frameworks (PAFs) as model adsorbents to systematically characterize this approach. A variety of material characterizations demonstrate that various PAF analogs can be successfully incorporated into polymer matrices with high dispersibility, compatibility, and robustness. Adsorption studies also show that PAFs maintain rapid, selective, and high-capacity ion binding capabilities within the membranes over at least 10 recycles. When placed into ion-capture electrodialysis setups, the composite membranes successfully capture Hg^{2+} from diverse Hg^{2+}-spiked nontraditional water sources while permeating all other competing cations, enabling simultaneous desalination. Finally, using other solute-selective PAF materials, we show that this strategy can in principle be applied generally to capture various target solutes.

11:30 am (Room Palm AD, Wednesday)

Oral 125 - Engineering subnanometer vacancy defects in single-layer graphene for high-performance gas separations

Luis Francisco Villalobos (Yale University)*; Kumar Varoon Agrawal (Ecole polytechnique fédérale de Lausanne)

Nanoporous single-layer graphene membranes, prepared by incorporating subnanometer vacancy defects in the graphene lattice, are highly promising for high flux gas separations because the resistance to diffuse is controlled by a single transition state at the nanopore. The main challenge for their realization is incorporating a high density of nanopores with the needed subangstrom resolution for relevant gas separations (e.g., 0.3-0.5 Å for separating CO_2 from N_2 and CH_4). Nanopores can be incorporated in the graphene lattice directly during its crystallization (intrinsic defects) or via post-synthetic etching strategies. In this presentation, I will make a critical comparison of these two routes, discuss the origin of the intrinsic defects, and share our recent success in boosting the density of gas-selective intrinsic defects to record densities (and hence performance).

I will discuss the advantages and disadvantages of bottom-up and post-synthetic strategies to engineer nanopores in graphene using, as examples, technologies developed in our lab. For example, a post-synthetic ozone etching chemistry that allows controlled nucleation and expansion of vacancy defects with a subangstrom resolution, and a controlled synthesis of nanocrystalline graphene where incomplete growth of nanometer-sized, misoriented grains generates molecular-sized pores in the lattice. Generally, post-synthetic strategies are believed to allow more control over the vacancy defects, and bottom-up approaches are considered more scalable. However, the answer might not be so definitive. I will share our recent results that show the scale-up potential of post-synthetic approaches. And how by understanding the origin of the intrinsic defects, we can engineer a strategy to control their density and size to produce high-performance gas-sieving membranes (e.g., a high density of gas-sieving pores that result in membranes with CO_2 permeance of 18,500 GPU and a CO_2/N_2 selectivity of 28).

12:00 pm (Room Palm AD, Wednesday)

Oral 126 - Designing Cation-Exchange Membranes with Coordination-Based Selectivity Between Cations

Ryan DuChanois (Yale University)*; Lauren A Mazurowski (Yale University); Menachem Elimelech (Yale University)

Ion separations are critical in energy production and water treatment technologies such as fuel cells, redox flow batteries, and electrodialysis. However, state-of-the-art polymeric membranes cannot attain the highly precise ion separations needed to meet emerging sustainability objectives, such as recovering valuable metals from wastewater to facilitate a circular resource economy. To leverage polymeric materials for metal recovery from wastewater, selective coordination sites may be needed within a membrane to provide molecular recognition.

In this talk, I will demonstrate metal recovery from multi-component feedwaters using cation-exchange membranes (CEMs) with high sorption selectivity for Cu^{2+}. A poly[(N,N-dicarboxymethyl) allylamine] (PDCMAA) Cu^{2+}-binding agent was synthesized via carboxymethylation of poly(allylamine) (PAH), and then commercial CEMs were modified with a thin film (~50 nm) of PAH and PDCMAA. In multi-component diffusion experiments, this surface modification increases the CEM selectivity toward Cu^{2+} over Mg^{2+} by a factor of ~20. The CEM maintains Cu^{2+} selectivity when the Mg^{2+} concentration exceeds the Cu^{2+} concentration 10-fold. We then project the performance of the individual components of the membrane (i.e., thin film and commercial CEM) and find the underlying CEM severely limits the separation performance of the composite membrane. Finally, I will propose how membrane design, including tailoring interactions between species of interest and an ultrathin selective layer, may lead to high precision and continuous ion-separation processes with polymeric membranes.

Oral 22 – MEMBRANE FOULING I

9:30 am (Room Palm CF, Wednesday)

Oral 127 - A Computational and Experimental Test Bed for Prediction of RO Module Fouling

Daniel J Miller (Lawrence Berkeley National Laboratory)*; Sergi Molins Rafa (Lawrence Berkeley National Laboratory); Mostafa Nassr (UT Austin); Sarah Dischinger (Lawrence Berkeley National Laboratory); Nicolas Spycher (LBNL); Jacob Johnston (Colorado School of Mines); Ji Yeon Lee (Lawrence Berkeley National Laboratory); Pedram Bigdelou (Colorado School of Mines); Kristofer Gleason (The University of Texas at Austin); Nils Tilton (Colorado School of Mines); William Stringfellow (Lawrence Berkeley National Laboratory); Benny Freeman (The University of Texas at Austin)

Fouling is a pervasive problem limiting the productivity of RO systems, and is typically addressed with a reactive (rather than proactive) approach (e.g., cleaning once flux drops below a threshold value). Changes in process conditions, such as variations in feed water quality (e.g., salinity, organic matter content), hydrodynamics (e.g., pressure, flowrates), and recovery frustrate efforts to control fouling. This presentation will describe efforts to develop computational models simulating feed flows in spiral-wound RO modules, including the impacts of channel geometry, feed spacers, and scaling. We leverage previous Department of Energy investments in geochemistry models, which have been developed to understand reactive transport processes occurring in brine flows through subsurface media. These models are informed by computational fluid dynamics calculations describing 2D fluid flow around feed spacers and 3D fluid velocity/salt concentration profiles in the feed channel. Computational models were able to predict the water permeate flux for commercial reverse osmosis membranes with and without feed spacers, and were extended to describe coupled reaction-transport phenomena at an RO membrane surface, including growth of scale and unsteady flow effects (e.g., vortex shedding) resulting from feed spacer geometry. Models are validated by laboratory crossflow fouling experiments that mimic the geometry of spiral-wound RO modules using commercial membranes under realistic flow conditions for feed waters containing single salts (e.g., $CaSO_4$) and complex salt mixtures. Experimental crossflow fouling experiments were performed in both constant flux and constant transmembrane pressure modes. To understand the effects of spacer geometry, 3D-printed feed spacers were studied in crossflow fouling experiments. Analytical techniques were developed for post-mortem autopsy of mineral scale on membrane surfaces.

10:00 am (Room Palm CF, Wednesday)

Oral 128 - Application of machine learning-based models to understand and predict critical flux of oil-in-water emulsion in crossflow microfiltration

Henry J Tanudjaja (Nanyang Technological University)*; Jia Wei Chew (Nanyang)

Random Forest (RF) and Neural Network (NN) were employed to respectively understand and predict the critical flux (Jcrit) of oil-in-water emulsions in crossflow microfiltration. A total of 223 datasets from various studies were compiled, with 9 operational parameters and one target variable of critical flux. RF indicates crossflow velocity (CFV) as the most dominant parameter in determining critical flux, outweighing surfactant and oil variations. Exceptions are found in specific cases when casein concentration was the most dominant, since the smaller sizes of casein significantly decreased Jcrit. The NN model predicted the best when all 9 input parameters were integrated, and worst when CFV was the sole parameter used for model development, even though CFV was identified as the most dominant. Results here demonstrate the usefulness of machine learning tools to enhance the understanding on and prediction of critical flux, without the need for any governing equations.

10:30 am (Room Palm CF, Wednesday)

Oral 129 - Quantifying Electrically Responsive Membranes for Biofouling Mitigation

Charles-François de Lannoy (McMaster)*; Amin Halali (McMaster University)

Electrically conductive membranes (ECMs) self-induce antifouling mechanisms at their surface under certain applied electrical currents. Quantifying these mechanisms is critical to enhancing ECMs' self-cleaning performance. Local pH change and H_2O_2 production are among the most important self-cleaning mechanisms previously hypothesized for ECMs, however the impacts of these mechanisms have not previously been isolated and comprehensively studied. We investigated antifouling mechanisms using CNT-based ECMs. We quantified the impact of electrochemically-induced acidic and alkaline conditions, and H_2O_2 concentration on, Escherichia Coli. We quantified the electrochemical potential of these ECMs to generate stressors (protons, hydroxyl ions, H_2O_2), under a range of applied currents (± 0-150 mA, 0 – 2.7 V). Next, these chemical stressors with similar magnitude to that generated at ECM surfaces were imposed on E. coli capsids and biofilms. We compared capsid and biofilm viability under these bulk and flow-through electrically conductive membrane-induced conditions, indicating significant similarities in the effects of extreme pH (40±10% viability) and significant difference in the effect of H_2O_2 (75 ±5% viability) exposure as compared with ECM-induced anti-fouling under 2 V applied potential. This study identified anti-biofouling mechanisms of ECMs and provides guidance for determining the current patterns that maximize their antifouling effects. Further, we propose a set of standardized characterization and performance tests to serve as a basis of comparison for ECM performance.

11:00 am (Room Palm CF, Wednesday)

Oral 130 - Feed Temperature Effects on Organic Fouling of Reverse Osmosis Membranes: Competition of Interfacial and Transport Properties

Santiago Romero-Vargas Castrillon (University of Edinburgh)*

Understanding the effect of feed temperature on fouling is essential for climate adaptation of water treatment infrastructure. We investigated the effect of feedwater temperature on the organic fouling of reverse osmosis (RO) membranes. Experiments were conducted over the range $27 \leq T \leq 40$ °C, relevant to feed temperatures in arid, near-equatorial latitudes. Fouling by alginate, a component of extracellular polymeric substances, was investigated at the nanoscale by AFM-based temperature-controlled colloidal-probe force spectroscopy (CPFS). The CPFS results, complemented by interfacial property characterization (contact angle, surface roughness, and charge) conducted under temperature-controlled conditions, enabled us to rationalize the observed fouling kinetics in cross-flow experiments. We observed less severe flux loss at 35 °C ($J/J0 = 75\%$, $t = 24$ h) compared to 27 °C ($J/J0 = 65\%$), which is consistent with weaker adhesion forces with rising temperature. The observed variation in the magnitude of adhesion forces agrees with the temperature dependence of hydrophobic interactions. At 40 °C, the observed flux loss ($J/J0 = 68\%$) was similar to that at 27 °C, despite the fact that adhesion forces are relatively weak (and similar to those at 35 °C). Analysis using a series-resistance model shows that the foulant hydraulic resistance is equal at 35 and 40 °C, consistent with the CPFS results. More severe fouling was observed at 40 °C compared to 35 °C, however, due to the higher water permeance at 40 °C, which resulted in a greater flux of foulant to the membrane. We further show that the fouling layer develops within ~2 h, during which the flux sharply decreases by 26% at 27 °C, 19% at 35 °C, and 22% at 40 °C; thereafter, flux losses are small and temperature independent. CPFS experiments show that this behavior is due to the foulant layer, which results in weak, often repulsive, and T-independent foulant– foulant interactions, which hinder further foulant deposition.

11:30 am (Room Palm CF, Wednesday)

Oral 131 - Acoustically excited microstructure for on-demand fouling mitigation in a microfluidic membrane filtration device

Kieran R Fung (Membrane Science, Engineering, and Technology (MAST) Center, Department of Mechanical Engineering, University of Colorado Boulder)*; Yuekang Li (Membrane Science, Engineering, and Technology (MAST) Center, Department of Mechanical Engineering, University of Colorado Boulder); Shouhong Fan (Membrane Science, Engineering, and Technology (MAST) Center, Department of Mechanical Engineering, University of Colorado Boulder); Apresio Fajrial (Membrane Science, Engineering, and Technology (MAST) Center, Department of Mechanical Engineering, University of Colorado Boulder); Yifu Ding (Membrane Science, Engineering, and Technology (MAST) Center, Department of Mechanical Engineering, University of Colorado Boulder); Xiaoyun Ding (Biomedical Engineering Program, Department of Mechanical Engineering, University of Colorado Boulder)

Over the past decade, acoustic streaming has been drawing increased interest in the field of acoustofluidics (integration of acoustics into microfluidic devices), for its rapid ability to disrupt laminar flow and manipulate particles. The fundamental mechanism used in acoustic streaming employs mechanical perturbations to oscillate microscale objects (structures or bubbles) to generate steady microstreaming effects within the flow field. Moreso, acoustic streaming effects are known to intensify mass transport around the microscale object and exert strong, localized shear forces. Capitalizing on these effects, we have developed a novel acoustofluidic membrane separation (AMS) device that employs acoustic streaming for rapid, on-demand membrane cleaning during particle separation operation. Our initial results indicate that acoustically excited microstructures can remove and breakup cake layer fouling within 100 milliseconds of acoustic excitation.

In this presentation, we experimentally probe acoustically excited microstructures and evaluate the ability to act as a novel physical membrane cleaning method. Coupled with microscopy and imaging techniques, acoustic streaming micromixing experiments are visualized and validated with flow profiles generated by our computation fluid dynamics (CFD) model. Following this, crossflow microfiltration experiments demonstrate that acoustically excited microstructures can effectively remove cake layer particle aggregations from the membrane surface and promote particle permeation across the membrane upon acoustic excitation.

12:00 pm (Room Palm CF, Wednesday)

Oral 132 - Unraveling the mechanism of an AC electric field applied for fouling mitigation

Saikat Bhattacharjee (Technion-Israel Institute Of Technology)*; Guy Z Ramon (Technion - Israel Institute of Technology)

Fouling is still a major challenge in membrane separation processes, adversely affecting system performance and, consequently, cost. Much effort has been extended towards controlling fouling, including passive methods such as various forms of surface modification, and active methods such as manipulating hydrodynamic conditions or applying an external electric field. Of the latter, DC electric fields have received the bulk attention, while the application of AC electric fields is far less understood.

In the present work, a theoretical model is derived to examine the effect of an applied AC field on the forces acting on a colloidal particle adjacent to the membrane surface. Recent work has shown that asymmetry, inherent to the transport of electrolytes due to the difference in diffusivity and charge of the ionic species present, can result in a non-zero, time-averaged electric field. This field persists over length-scales far greater than the electric double layer thickness, and will present a long-range electric repulsion force on a like-charged particle. The developed framework extends the existing theory by including the effect of permeation through a membrane. Specifically, we seek the parametric phase-space permitting equilibrium positions to exist, whereby a particle, dragged toward the membrane by permeation, is then held fixed, at a finite distance from the membrane, through the action of the time-averaged electric force. The dependence of such equilibria on system parameters is explored, revealing the regime where AC fields may be applied for colloidal fouling mitigation.

Oral 23 – MEMBRANE DISTILLATION AND PERVAPORATION

9:30 am (Room Abbey North, Wednesday)

Oral 133 - Vapor Flux of Membrane Distillation: Theoretical Limits, Insights on Membrane Design, and Anomalous Phenomena

Ruoyu Wang (vanderbilt university)*; Shihong Lin (Vanderbilt University)

Numerous MD studies are published in recent years with different membrane types, varying from polymeric membranes to nanomaterial modified or embedded membranes. While increasing effort has been devoted to designing membrane with resistance to fouling, wetting, and scaling, many studies on MD membrane design were still motivated to achieve a high flux. Desired membrane properties to enhance flux are larger pore size, higher porosity, lower tortuosity, and smaller thickness, etc. Typical vapor flux in DCMD varies from several to tens of LMH depending on membrane properties and operating conditions. However, some recent works have reported anomalous high flux of hundreds of LMH. Unfortunately, a fair comparison between membranes from different studies is challenging as different MD configurations and operating conditions are applied. To better evaluate the performance enhancement of newly designed MD membranes and guide future development, it is necessary to explore the theoretical MD vapor flux limit using the well-established mass and heat transfer model.

In this presentation, we first explore the theoretical flux limit of symmetric single-layer MD membranes by hypothetically increase membrane porosity to approach unity and increase pore size to infinity, assuming the feed and distillate are still physically separated by a fictional "magic air gap (MAG)". Then, we explore the performance limit of composite membranes (i.e., a thin hydrophobic layer and a thicker hydrophilic support layer) by replacing the hydrophobic layer with an MAG and investigate the impacts of the thickness of the hydrophilic support layer and the membrane orientation. Additionally, we analyze the anomalous high fluxes from literature with reported membrane properties and show they have exceeded the theoretical limit. Possible mechanisms for anomalous high flux will be discussed, including the increase of effective evaporation area and faster evaporation from nanopores.

10:00 am (Room Abbey North, Wednesday)

Oral 134 - Modeling Pilot-Scale Membrane Distillation at High Salinity using Bench-Scale Data

Mukta Hardikar (The University of Arizona)*; Itzel Marquez (Central Michigan University); Eduardo Saez (University of Arizona); Andrea Achilli (The University of Arizona)

Membrane distillation (MD) is a promising technology for treating high concentration wastewater. However, at pilot-scale, MD operates at a smaller transmembrane temperature difference (than bench-scale) because of a longer membrane, resulting in a smaller driving force for water flux. Modeling water flux in pilot-scale MD at high salinity can help determine operating conditions (feed flow rate, inlet temperature) and membrane element geometry that maximizes water production and minimizes specific thermal energy consumption (STEC). In this work, we developed a procedure to use bench-scale direct contact MD (DCMD) experimental data to model pilot-scale air-gap MD (AGMD). DCMD is a commonly used configuration at the bench-scale because of its simplicity in design while AGMD is commonly used at pilot-scale because of better energy efficiency. Bench-scale experimental data was used to determine membrane permeability on two different membranes, a polyterafluoethylene (PTFE) and a polyethylene (PE) membrane. A stepwise modeling approach was then used along with local energy and mass balance to estimate temperature and concentration polarization along the pilot-scale element and predict water flux. The model was validated with pilot-scale experimental data obtained over a wide range of operating conditions (feed flow rate - 2 to 15 L/min - temperature – 40 to 80 °C, and salinity – 0 to 200 g/kg). The validated model was used to identify effect of flow rate, temperature, element geometry and thus residence time in the membrane element on water production and STEC based on feed salinity. Interestingly, at a given salinity, a minimum STEC exists and was observed to be a function membrane element geometry and flow rate and did not always overlap with maximum water flux. Thus, the model can act as a tool to decide operating conditions or geometry of membrane elements based on feed salinity and available heat to maximize either water flux or energy efficiency.

10:30 am (Room Abbey North, Wednesday)

Oral 135 - Integrated electrocoagulation, membrane filtration and membrane distillation for treatment of hydraulic fracturing produced water

Ranil Wickramasinghe (University of Arkansas)*; Mahmood Jebur (University of Arkansas); YuHsuan Chiao

Membrane based separations are attractive for water and wastewater treatment. Membrane based unit operations are environmentally friendly, easily scaled up, have a small footprint and generally have low operating costs. Further unit operations such as submerged membranes in membrane bioreactors offer significant process intensification compared to conventional activated sludge processes. However, membrane fouling is a major impediment which limits the viability of membrane based unit operations. Consequently, integration of a membrane based unit operation with appropriate feed pretreatment is critical to minimize membrane fouling. In addition, development of advanced fouling resistant membranes is essential.

In this presentation the use of electrocoagulation as a feed pretreatment operation prior to membrane distillation will be described. The advantages of an integrated electrocoagulation, microfiltration and membrane distillation process for maximizing water recovery from hydraulic fracturing produced water will further highlight the potential for process intensification through integrated membrane based separation processes.

While these examples highlight the importance of developing integrated membrane based separation processes to minimize membrane fouling and maximize water recovery the need to develop advanced fouling resistant membranes persists. For example, a major challenge that limits the viability of membrane distillation is fouling and wetting of the membrane. Membrane fouling can lead to a significant decline in the permeate flux while wetting leads to passage of water rather than the water vapour through the membrane pores. Novel bilayer electrospun membranes that are fouling and wetting resistant will be described. We show that tuning the surface properties of the membrane surface that faces the feed and permeate streams is essential to increase the resistance to fouling and wetting of the membrane.

11:00 am (Room Abbey North, Wednesday)

Oral 136 - Fermentation broth recovery of 2,3 butanediol using polymer based ceramic membrane for biofuel production.

Oindrila Gupta (Oak Ridge National Laboratory)*; Aimee Church (Oak Ridge National Laboratory)

Rapid industrialization over the last few decades have resulted in shortage of natural resources with energy needs rising exponentially. Simultaneously, there is a tremendous increase in greenhouse gas emissions as fossil fuels still constitute about 85% of the world's energy demands. Production of biomass derived fuels can be a sustainable alternative. To solve this issue, research initiatives have focused on microbial fermentation for production of platform chemicals. One such versatile bulk chemical produced via bacterial fermentation is 2,3-butanediol (2,3-BDO). However, its recovery/separation has several challenges in catalytic upgrading to fuels owing to its low content, high-water solubility and high boiling point along with complex fermentation compositions. Conventional approaches involve high pressure steam recovery when distillation is implemented, that is expensive and not environmentally friendly. In this work, we have incorporated pervaporation, a membrane-based technology to concentrate 2,3 BDO from fermentation broth. A novel membrane – polymer coated ceramic membrane has been used to combine the advantages of both ceramics and polymeric materials for improved membrane performance and eventually achieve better separation at a reduced cost with more energy savings. Our polymer-coated ceramic membrane allowed for selective high-throughput permeation of water and selective removal of impurities at low to moderate temperatures (60-80°C). Compared to traditional ceramic or polymeric membranes, our polymer-coated ceramic membranes offered the potential of reduced swelling and provided better mechanical ruggedness and pervaporation stability over extended time periods. These results are promising for large scale 2,3 BDO recovery from fermentation broth, making the approach environmentally sustainable.

11:30 am (Room Abbey North, Wednesday)

Oral 137 - Impact of oxidative chemicals on the performance and materials properties of hydrophobic porous membranes used in membrane distillation

Hannah Cairney (University Of Colorado Boulder); Elizabeth A Hjelvik (University of Colorado Boulder)*; Duong T. Nguyen (University of Colorado Boulder); Anthony Straub (University of Colorado Boulder)

Hydrophobic porous membranes, such as those used in membrane distillation (MD), are increasingly implemented for advanced water treatment. Despite growing use of these membranes, their interactions with chemical oxidants commonly used in water treatment are poorly understood. Understanding the resistance of these membranes to oxidative degradation will provide opportunities to combine oxidation and membrane-based treatment for reduced fouling and enhanced contaminant removal. In this work, we determine how exposure to oxidants commonly used in water treatment will affect the overall chemical structure and subsequent performance of hydrophobic membranes. To probe oxidative behavior, polytetrafluoroethylene (PTFE), polyvinylidene fluoride (PVDF), and polypropylene (PP) membranes were exposed to chlorine and ozone at varying exposure doses at a pH of 4, 7, and 10. The extent of degradation for each polymer was determined using water contact angle measurements, fourier transform infrared spectroscopy (FTIR), X-ray photoelectron spectroscopy (XPS), and scanning electron microscopy (SEM). While water flux varied with oxidant exposure, there was no significant decrease in salt rejection observed regardless of the oxidant or exposure time for PTFE and PVDF membranes. PP membranes showed mechanical failure that compromised membrane performance at high oxidant exposures . The results of this work indicate a high resistivity of PTFE and PVDF membranes to degradation by conventional oxidants.

12:00 pm (Room Abbey North, Wednesday)

Oral 138 - Porous Metal Foam Condensation Doubles Energy Efficiency of Membrane Distillation

Hamid Fattahijuybari (Purdue University)*; Harsharaj B. Parmar (Purdue University); Ali D. Alshubbar (Purdue University); Katherine L. Young (Purdue University); David M Warsinger (Purdue)

Membrane Distillation (MD) is a thermally driven desalination process that uses the vapor pressure difference across a hydrophobic membrane as a driving force to extract pure water from varied working mediums. The benefits of the membrane distillation process over conventional RO systems include handling very high feed salinity, using simple membrane characteristics, and easy coupling with renewable and waste heat sources. Purifying water with membrane distillation from saline feedstocks is energy-intensive and costly, especially at high salinities. This requires improving heat and mass transport in membrane distillation systems to operate efficiently. These systems closely resemble heat exchangers, which use that design for internal latent heat recovery by vapor condensation. Porous metallic condensers can substantially improve condensation heat transfer and are a well-known way to improve the heat transfer in the heat exchangers. Metal foams have been extensively studied in heat exchangers to achieve 2-3 times better heat transfer coefficients. In the present work, we use a porous metal foam in the air gap of air gap membrane distillation (AGMD) to increase the overall energy efficiency of water flux and mitigate the flooding of the air gap. MD experiments show that the porous copper foam improves the thermal efficiency of air gap membrane distillation up to 40% and water flux up to 144%. Numerical modeling shows the porous copper foam improves the Gain output ratio (GOR) of membrane distillation in high salinity by 98.7%.

Oral 24 – INDUSTRIAL APPLICATIONS

9:30 am (Room Palm BE, Wednesday)

Oral 139 - An Enterprise Approach to Developing Industrial Membrane-based Solutions

Adil M Dhalla (Separation Technologies Applied Research and Translation (START) Centre, Nanyang Technological University - NTUitive Pte Ltd)*; Chakravarthy Gudipati (START - NTUitive Pte Ltd); Joanne Sim (SG MEM, Singapore Membrane Consortium, Nanyang Technological University - NTUitive Pte Ltd)

One of the biggest challenges for commercialization of novel ideas, even if the Intellectual Property is duly protected, is the gap between laboratory processes, results and testing, and the full scale final product. Key risks include scale-up of component materials and equipment, systems level thinking, testing at pilot scale in an actual application setting, and final implementation.

Singapore's Membrane Consortium, SG MEM, was set up to enable partnerships and collaborations towards developing Platform Solutions across our Membrane Ecosystem. It brings together early stage research from our universities, Singapore's unique translational facilities, and industry partners from upstream (materials companies), to membrane manufacturers, solution providers and end-users of separations technologies. This expanding and varied group of companies ranges from start-ups to SMEs, large local enterprises to multinationals.

One of the key institutional members of this ecosystem is the Separation Technologies Applied Research and Translation (START) Centre, Singapore's national facility for bridging the gap between promising innovations in separations, especially membrane based inventions, at the laboratory scale, and industrial scale products and processes. Over the past three years, this centre has built up broad capabilities in membrane (both flat-sheet and hollow-fiber) fabrication at industrial scale, the design, construction and testing of elements and modules, and the design of pilot systems for testing in real-life scenarios.

This talk will showcase how we have built the framework to take early stage membrane inventions to commercially viable solutions for key challenges in the fields of Water and Environment. Examples include technologies focused on industrial waste-water treatment for re-use, including potential recovery of valuables from the waste stream, and unique low-energy solution for water softening, both with multiple applications.

10:00 am (Room Palm BE, Wednesday)

Oral 140 - Single Pass TFF for Monoclonal Antibody Processing – Advancing a new technology through the MAST Center

Andrew Zydney (Penn State University)*

There is growing interest in the development of integrated continuous bioprocesses for the production of important biotherapeutics due to the potential for enhanced productivity, greater flexibility, and improved product quality. Single pass tangential flow filtration (SPTFF) can greatly facilitate this transition, providing inline concentration and buffer exchange in the production of monoclonal antibodies that are used for the treatment of autoimmune diseases, cancer, and Covid-19. However, the design and optimization of SPTFF modules can be quite challenging due to variations in flow, pressure, and antibody concentration throughout these long pathlength modules. This collaborative project, sponsored by the MAST Center, was focused on the development of a new mathematical model that could predict the flux / pressure profiles in SPTFF and then optimize the design of SPTFF processes for specific target applications. Model validation was performed using a monoclonal antibody feed-stream provided by Biogen, with experimental data obtained in SPTFF modules provided by Pall and Millipore, all industrial members of the MAST Center. The model demonstrated that the greatest concentration factor could be achieved using a long thin channel, although this resulted in very large feed-side pressure drops. The use of cascade configurations, with a greater number of parallel channels near the feed inlet, significantly reducing the pressure drop but with a corresponding increase in total membrane area needed for the inline concentration. These results not only provide important insights into the design and optimization of SPTFF systems for monoclonal antibody processing, they also demonstrate the unique capabilities of this type of university-industry consortium in helping advance new membrane technologies for the production of high value biopharmaceuticals.

10:30 am (Room Palm BE, Wednesday)

Oral 141 - Membranes Beyond Water - Deploying Spirals in Industrial Separations

Benjamin Weaver (Solecta)*

Spiral wound elements have been utilized for water treatment from applications such as desalination to waste water recovery. However, there are many novel membrane applications that go beyond water. This presentation will discuss the application of MF, UF, NF and RO membranes in industrial applications that has enabled industry to solve difficult process challenges by deploying membrane technology. In the dairy industry membranes are applied to separate and concentrate whey protein has allowed to turn a waste product into a revenue stream. The importance of sanitization and element design will be discussed. We will then move to the automotive world where the utilization of ultrafiltration membranes in a novel configuration have allowed for an economical painting process and utilizing NF/RO for reduction of water usage. Finally in the world of food ingredients and nutraceuticals, the boundaries of pushing higher solids and higher viscosity solutions through spiral wound elements using different element configurations to process fluids that historically used alternative technologies for separations will be discussed. Throughout the discussion we will tie together the importance of membrane selection, element and module configuration and design, and application development in industrial processes.

11:00 am (Room Palm BE, Wednesday)

Oral 142 - Separation and Recovery of Critical Materials from End-of-Life Lithium Ion Batteries using Membrane Solvent Extraction Process

Syed Islam (Oak Ridge National Laboratory)*; Priyesh Wagh (Oak Ridge National Laboratory); Ramesh Bhave (Ornl); John Klaehn (Idaho National Laboratory)

In recent years, LIBs have drawn huge attention worldwide due to their wide use in hybrid/electric vehicles and portable electronics. LIBs contain many critical elements including cobalt, nickel, and lithium. Although the demand for LIBs is growing rapidly, the production of those critical elements is limited in mining. To address the growing demand, end-of-life LIBs could be a potential secondary source of the critical materials. However, currently, there are no established technologies for the recycling of end-of-life LIBs to meet the growing market demands of the critical materials. Currently, only less than 5% of LIBs are recycled due to high processing cost and inefficiency of the existing technologies. In order to achieve the above goal, we have developed a highly energy-efficient, cost-effective, and environmentally friendly membrane solvent extraction (MSX) process for the recovery of high purity critical elements from spent LIBs. The MSX process has been demonstrated for separation and recovery of critical metals including cobalt, nickel and lithium with >99.5% purity and yield of >90% from a wide range of end-of-life LIBs. MSX technology is based on the concept of utilizing low-cost robust microporous membrane supports for solvent extraction using a highly selective extractant. Utilizing hollow fiber modules, the MSX system provides a compact and modular configuration, achieving high extraction rates for desirable elements with low energy and cost requirements from a wide variety of feedstocks. Additionally, the chemical inventory, especially the volume requirements of organic phase in MSX process are substantially low compared to the traditional solvent extraction, which can help with lower operating cost. MSX has the potential to address the challenges to secure a domestic source of critical materials from LIBs for reuse in new LIB formulations. This novel process supports a closed-loop circular economy for clean energy technologies.

11:30 am (Room Palm BE, Wednesday)

Oral 143 - Industrially relevant academic research – Answering fundamental questions at the module scale

Jeffrey R McCutcheon (University of Connecticut)*; Noah Ferguson (University of Connecticut); Maqsud Chowdhury (University of Connecticut); Johann Schnittger (Fraunhofer-Institut für Keramische Technologien und Systeme IKTS); Marianne Pemberton (University of Connecticut); Edward Wazer (University of Connecticut); Mayur Ostwal (University of Connecticut); Marcus Weyd (Fraunhofer-Institut für Keramische Technologien und Systeme IKTS); Joachim Koschikowski (Fraunhofer-Institute for Solar Energy Systems ISE)

Academic-industrial partnerships are particularly valuable when fundamental research questions can be explored at scale or with relevant applications. In this work, we describe two of our industrially relevant projects. These projects represent the exploration of fundamental research questions using module-scale membranes.

The first is an exploration of forward osmosis (FO) at module scale. This work shows how FO systems perform when placed in tandem with reverse osmosis (RO) for draw solution regeneration. The module-scale testing provides insight into the self-regulating behavior of FORO systems and how they can be operated with minimum control. We evaluate full-scale modules from Fluid Technology Solutions and Aquaporin.

Our second highlighted project is on solar-driven desalination. This work considers the use of ceramic membranes for high temperature membrane distillation for the concentration of brines. Our work explores single and multi-channel ceramic membrane elements to evaluate their performance under varying conditions.

12:00 pm (Room Palm BE, Wednesday)

Oral 144 - Facilitated Transport Membranes for Low Energy and Low Cost Carbon Capture and Sequestration

Christine Parrish (Compact Membrane Systems)*; Sudip Majumdar (Compact Membrane Systems); Ken Loprete (Compact Membrane Systems)

With the sustainable energy transition underway, it is vital to reduce greenhouse gas emissions from existing processes in the short term to continue to produce reliable and affordable power. A majority of the world still relies on fossil fuel technology for power, and coal-fired power plants account for 30% of global carbon dioxide (CO_2) emissions. Carbon capture technology allows users to filter and sequester the CO_2 at the source and reduce emissions by up to 90%. Currently, separation and concentration of the CO_2 stream in preparation for liquefaction and storage is done using amine towers. These towers are costly (>$50/ton of CO_2) and have a parasitic energy load of 20-40%, making them unfeasible for large scale adoption. Membranes hold promise for carbon capture because they can be scaled to the size of the application and require lower energy usage and operational cost than existing carbon capture technologies.

Optiperm™ carbon is a facilitated transport membrane with significantly higher permeance than existing membrane technology. The technology uses a fluoropolymer composite membrane with a facilitating agent incorporated into the polymer to pull CO_2 across the membrane. The increased flux and high selectivity positions Optiperm™ carbon to reduce the cost of carbon capture across multiple application. Membranes operate in energy efficient modes, so can reduce the parasitic load imposed by alternative technologies like amine towers.

Initial results for utilizing the CMS membranes to capture carbon dioxide from flue gas, including the effect of operating conditions and initial system modeling, will be shared. Results thus far have shown a very high flux membrane (2000-5000 GPU) that would result in substantially lower capture costs and increased efficiency.

Oral 25 – MATERIALS FOR ELECTROCHEMICAL APPLICATIONS

9:30 am (Room Abbey South, Wednesday)

Oral 145 - Pressure Induced Diffusion in Hydrated Polymers

Rahul Sujanani (The University of Texas at Austin)*; Kevin Reimund (University of Texas at Austin); Kris Gleason (The University of Texas at Austin); Benny Freeman (The University of Texas at Austin)

Water transport in hydrated polymers has been studied in considerable detail in the context of various membrane materials (e.g., reverse osmosis, ion exchange, nanofiltration, etc.). However, there are conflicting reports about the fundamental physics of pressure driven water transport in such materials, with some authors postulating a pore flow mechanism and others applying the solution/diffusion model. For example, researchers studying ionomers have proposed a heterogenous pore structure in hydrated Nafion, speculating that a pressure gradient through the thickness of the membrane, rather than a concentration gradient, drives water transport in operations where the feed is pressurized. However, to the best of our knowledge, there is no evidence of fixed, permanent pores in Nafion. To address this topic, this presentation will detail experimental measurements of pressure driven water transport in Nafion and related materials. Inspired by the seminal work of Rosenbaum and Cotton, several films were stacked together in a permeation cell, with the feed side pressurized between 1000-2500 psi and the permeate side left at atmospheric conditions. At steady state, the films were separated, and the water content of each film was measured, revealing a water concentration gradient through the membrane stack that scales with feed pressure: a distinct feature of the solution-diffusion mechanism. The thermodynamic origins of this behavior will be discussed. In general, these results are consistent with the historical literature detailing pressure driven liquid transport in dense (i.e., non-porous) membranes, connecting teachings from several communities interested in transport through membranes.

10:00 am (Room Abbey South, Wednesday)

Oral 146 - Kinetic control of intrinsic pores in monolayer graphene for large-area proton selective membranes

Piran Kidambi (Vanderbilt University)*

Angstrom scale proton selective defects introduced into the lattice of atomically thin 2D materials such as graphene via scalable processes present potential for transformative advances as membranes for energy conversion/storage applications. Here, we show for the first time that kinetic control of graphene synthesis during chemical vapor deposition (CVD) can allow for tunable Angstrom scale proton selective pores for facile fabrication of large-area atomically thin proton exchange membranes. Using a resistance-based transport model in conjunction with systematic liquid and gas phase transport measurements on the same membrane, we study the influence of such Angstrom scale intrinsic defects on selective proton transport through centimeter-scale Nafion|Graphene|Nafion sandwich membranes. Additionally, we report on a novel graphene stacking approach that effectively mitigates transport of un-desired species(small ions and gas crossover)without adversely affecting proton transport. Our insights on kinetic control of graphene synthesis for Angstrom scale pore formation over large areas as well as facile stacking approach for membrane fabrication offer new avenues to enable functional large-area atomically thin proton exchange membranes.

10:30 am (Room Abbey South, Wednesday)

Oral 147 - Exploiting the synergistic effect of surfactants and ionic liquids for CO_2 capture

Alireza Bandegi (Case Western Reserve University); Maria Marquez Garcia (New Mexico State University); Jose Leobardo Bañuelos (The University of Texas at El Paso); Millicent Firestone (Los Alamos National Laboratory); Reza Foudazi (University of Oklahoma)*

Ionic Liquids (ILs) are promising candidates for carbon dioxide (CO_2) capture due to their high capacity and selectivity. Developing methods that enable the cost-effective and regenerable operation using ILs-based absorbent is important to carbon dioxide capture and storage (CCS) technology. In this study, we investigate the feasibility and performance of IL immobilized in lyotropic liquid crystal (LLC) ion gels for CO_2 absorption. The LLCs are obtained through self-assembly of a short chain block copolymer (BCP) of polyethylene-block-poly(ethylene oxide), PE-b-PEO, as surfactant in ILs. The results show a synergistic effect on the CO_2 absorption capacity of ion gels as a result of LLC structure and interaction of ILs with PEO blocks. In addition, the type of ions in ionic liquid significantly affect both the dynamics of PEO blocks in LLC ion gels and the final absorption capacity. The proper choice of composition for the LLC ion gels allows the immobilized ILs to outperform their liquid counterpart. This novel materials design allows for higher CO_2 absorption compared to neat IL and provides insight into improved carbon capture technologies.

11:00 am (Room Abbey South, Wednesday)

Oral 148 - Computational modeling of 3D electrospun nanofiber structures

Albert Kim (U. of Hawai'i)*

Electrospun nanofibers are generated using a process sharing characteristics of electrospraying and dry spinning, which have various applications such as filtration, textile and pharmaceutical manufacturing, catalysis, and medical/cosmetic delivery processes. Manufactured electrospun nanofibers have complicated 3D structures, which are strongly controlled by various operational parameters in three categories: first, viscosity, concentration, molecular weight, conductivity, and surface tension of polymer solutions; second, voltage, tip-to-collector distance, feed rate; and third, humidity and temperature. The electrospun nanofibers can be considered a case of densely packed, fibrous porous media. Due to the limit of measuring devices, only surface structures of the electrospun nanofibers can be imaged to investigate their structural characteristics visually. To understand the structural influences of the electrospun nanofibers, it is of great necessity to generate digital 3D structures of the nanofibers using simple algebraic models. This study shows the mathematical approaches to create 3D digital structures and predicts the material characteristics of the electrospun nanofibers.

11:30 am (Room Abbey South, Wednesday)

Oral 149 - Selective separation of lithium using chemically modified nanoporous polyamide membrane with electrodialysis

Xi Chen (Stanford University)*; William Tarpeh (Stanford University)

Current demand of lithium is expected to exceed global lithium reserves by 2040. While a vast amount of lithium exists in brines and wastewater, current technologies for mining and recycling lithium from the aqueous phase suffer from high energy intensities and inadequate selectivity of lithium from co-existing ions. Advancing solute-specific selectivities has also become the new frontier of membrane technology. In this study, we present a simple approach to chemically modify commercially available nanoporous polyamide membranes to enhance the density of carboxyl groups in the membrane matrix, thus achieving a ligand-enhanced membrane. The membrane employs the difference in solute-membrane affinities to distinguish lithium from other ions. Unlike pressure-driven processes that separate a large volume of water from solutes, the ligand-enhanced membrane is applied in the energy-efficient electrodialysis process, where only solutes are transported across the membrane. We challenged the ligand-enhanced membrane with lithium solution mixed with magnesium ions, which commonly exists in lithium brines. Our findings show that transport of magnesium is significantly facilitated compared with that of lithium, achieving a Mg/Li selectivity up to 4. Our results highlight that Li/Mg separation can be achieved by effectively concentrating Li while removing Mg. Densities of the carboxyl groups in the membrane are quantified and related to the selectivity performance. Influences of pH on the ionization/deionization behaviors of the membrane are examined. Impacts of the orientation of the membrane active layer are also investigated. This study shows a potential membrane-based method for aqueous separation of lithium, and also provides important insights into strategies for preparing membranes with solute-specific selectivities.

12:00 pm (Room Abbey South, Wednesday)

Oral 150 - Counter-ion Activation Energy of Diffusion in Ion Exchange Membranes

David Kitto (University of Michigan - Kamcev Lab)*; José Carlos Díaz (University of Michigan); Carolina Espinoza (University of Michigan); Jovan Kamcev (University of Michigan)

Understanding how to control ion transport in ion exchange membranes (IEMs) is a crucial task which will ultimately improve the performance of electrochemical applications employing these materials. Little is currently known about how to tune the membrane counter-ion selectivity because charge-exclusion principles, which govern ion sorption in IEMs, do not apply between like-charged species. Understanding the diffusive selectivity is therefore of great importance. To that end, the activation energy of counter-ion diffusion in an IEM is a useful metric for quantifying the fundamental mass transport resistance of an IEM. Despite being an uncommon characterization tool for IEMs intended for desalination, determining the activation energy of diffusion is a remarkably versatile method to understand the mechanism of ion transport at a molecular level, making such characterization valuable to investigators studying IEMs. To amend this near-omission in the broader literature, we measured the activation energy of diffusion for thirteen monovalent or divalent counter-ions in homogeneous ion exchange membranes equilibrated with deionized water using electrochemical impedance spectroscopy. One anion-exchange membrane and one cation-exchange membrane with comparable water contents and charge densities (measured in their Cl- and Na+ forms, respectively) were synthesized for this expansive study. The trends of counter-ion activation energy are analyzed and correlated with other intrinsic membrane and ionic properties, like the ionic size, ionic hydration energy, bond vibration energies (probed via Fourier transformation infrared spectroscopy), states of water in the membrane (probed via differential scanning calorimetry), water contents, and charge densities. The ultimate goal of this study is to develop an improved understanding of which key membrane and ion properties primarily impact the energy barrier of counter-ion transport across IEMs.

Abstracts – Oral Presentations, Wednesday

Presenting authors are indicated by an asterisk*. See the most up-to-date version of the program book with full abstracts at **www.membranes.org/nams~2022/**

Oral 26 – SEAWATER DESAL AND OSMOTIC PROCESSES

2:00 pm (Room Palm BE, Wednesday)

Oral 151 - Interfacial Polymerization kinetics: new insights on film formation using in-situ microscopy and particle-tracking

Adi Ben Zvi (technion)*; Guy Z Ramon (Technion - Israel Institute of Technology)

Thin-film polyamide (PA) selective layers fabricated via interfacial polymerization (IP) form the core of membrane-based water purification. Despite widespread commercial application, the exact details of reaction kinetics and film formation are not fully understood. Studying reaction kinetics is critical for controlling the IP reaction and designing better membranes for different applications. However, studies mostly rely on post-fabrication characterization due to the challenging nature of IP that makes it almost impossible to monitor. Recently, in-situ monitoring techniques have evolved, enabling direct monitoring of reaction kinetics in terms of heat and mass transfer and reaction dynamics.

Here, we study reaction kinetics and film formation via microscopy-based particle tracking performed during IP. Nanometer-scale fluorescent particles located in the aqueous phase are observed using confocal microscopy during an IP reaction occurring in a microfluidic device. Moreover, temperature-sensitive fluorescent dye is added to the aqueous phase to assess heat release during the reaction. Reaction parameters that affect the polymer formation rate and/or the system's susceptibility to instability are systematically varied, while particle motion and temperature are monitored. This enables identification of possible instability mechanisms during IP, which manifest as particle motion and interfacial deformation, that ultimately affect film formation and its resultant morphology. Furthermore, the duration of film formation as well as the evolution of its mechanical properties are evaluated. Such measurements are important steps in the process of acquiring fundamental understanding of IP and mechanisms that affect the evolvement of different film morphologies.

2:30 pm (Room Palm BE, Wednesday)

Oral 152 - A Solution-diffusion with Defects Model to Explain Pressure-dependent Membrane Permeability in Pressure Retarded Osmosis

Zachary Binger (The University of Arizona)*; Andrea Achilli (University of Arizona)

Physical deformation of the selective layer of membranes, affecting transport dynamics, has previously been the de-facto explanation for experimental results showing pressure-dependent membrane permeability. In this work an alternative model, validated with experimental data retrieved from an engineering-scale PRO system, is proposed. A solution-diffusion with defects (SDWD) model proposes that a membrane with defects, where the defects operate like a filtration membrane, would allow for a pressure-driven convective flow through the membrane. In the case of PRO, this will cause bulk transport of fluid from the draw stream into the feed stream effectively lowering permeate recovery and salt retention. An engineering-scale seawater reverse osmosis-pressure retarded osmosis (SWRO-PRO) system was designed and tested to evaluate the potential energy recovery of an integrated pressure retarded osmosis (PRO) system in desalination applications. During experimental testing, gross specific energy recoveries as high as 0.14 kWh/m^3 were achieved. However, energy consumption due to pressure losses in the system reduced the net specific energy recovery to a maximum of -0.07 kWh/m^3, indicating that permeate recovery and energy consumption are still challenges to PRO application. Furthermore, these experimental results indicated that the water and salt permeability of the PRO membrane varied greatly as the applied pressure changed which resulted in lower-than-expected permeate recovery. Modeling software utilizing the SDWD model was used in tandem with experimental data to demonstrate the presence and extent of defects in the experimental membrane structure, that allow for pressure-driven pore flow. The SDWD model explains the lower-than-expected water permeability and salt retention often seen in the experimental PRO results and integrating the SDWD model into future software will allow for more accurate simulation of larger systems and aid in investigations of PRO scale-up.

3:00 pm (Room Palm BE, Wednesday)

Oral 153 - Elucidating the Roles of Polyamide Layer Structural Properties in the Permeability-Selectivity Tradeoff Governing Aqueous Separations

Xi Chen (Stanford University)*; Rhea Verbeke (KU Leuven); Chanhee Boo (Korea Institute of Science and Technology); Marcel Dickmann (Universität der Bundeswehr München); Werner Egger (Universität der Bundeswehr München); Ivo Vankelecom (Universität der Bundeswehr München); Ngai Yin Yip (Columbia University)

Thin-film composite polyamide (TFC-PA) membranes are widely used in reverse osmosis and nanofiltration, and a tradeoff relationship between permeability-selectivity is extensively observed in these membranes. Fundamental understanding of the first principles governing the tradeoff is still incomplete. This study investigates the roles of key structural properties of the PA layer, i.e., size of free volume elements (FVEs) and effective permeant transport pathway, in the tradeoff behaviors. TFC-PA membranes with different selectivities are obtained from chemical modifications, and the permeability-selectivity trends are examined using neural organic solutes. Size of the FVEs of the polyamide layer are characterized using positron annihilation lifetime spectroscopy (PALS), and the effective transport pathway of membranes with different selectivities are analyzed with the hindered transport model. Our study shows that changes in the FVE sizes at rising membrane water permeability are not significant, and do not completely account for the water permeability trend. Impacts of physicochemical properties of the polyamide matrix on permeant transport are discussed, and our analysis further shows that the substantial shortening of the effective permeant transport pathway is a principal factor governing the water transport. Analysis using the hindered transport model reveals that selectivity behaviors are dominated by the effective PA pore size, and are relatively independent of transport pathway. Lastly, we reconcile the experimental results and theoretical model to propose a mechanistic framework to elaborate how structural properties of the polyamide layer influence the permeability-selectivity tradeoff. Findings of this study provide significant new insights for understanding transport mechanisms in all osmotically-driven membrane processes, and lays the foundation for rationally-informed design of next-generation membranes.

3:30 pm (Room Palm BE, Wednesday)

Oral 154 - The significance of membrane charge and co-ion partitioning in determining salt transport through reverse osmosis membranes

Li Wang (Yale University)*; Kevin Pataroque (Yale University); Tianchi Cao (Yale University); Menachem Elimelech (Yale University)

Polyamide membranes with a fully aromatic active layer are at the core of the reverse osmosis (RO) desalination process. RO performance is governed by the extent of salt and water permeation through the membrane active layer. The partitioning of salt ions into the polyamide active layer plays an important role in salt transport. A few studies employed electrochemical impedance spectroscopy (EIS) and quartz crystal microbalance (QCM) to investigate the partitioning of salt into the active layer. However, the estimated partitioning coefficients do not agree with the salt rejection observed in RO experiments. For example, the measured partition coefficient decreases as the salt concentration increases, while observed salt transport increases when RO feed salt concentration increases.

Recent work has demonstrated that the salt flux across polyamide RO membranes is controlled by the transport of co-ions (i.e., anions bearing an opposite charge to the negatively charged polyamide membrane). Herein, we develop a method to quantify the partitioning coefficient of co-ions by using QCM. We first evaluated the partitioning coefficient of co-ions as a function of salt concentration, revealing that co-ion partitioning increases at higher feed salt concentrations. We then assessed the effects of solution pH on the partitioning of co-ions. Our results demonstrate an increase of the partitioning coefficient of co-ions at lower solution pH, which we ascribe to the lower dissociation of carboxyl functional groups in the membrane matrix. The Donnan theory is employed to rationalize the dependence of co-ion partitioning on salt concentration and solution pH. In contrast to previous studies, the behavior of the co-ion partitioning coefficients at various salt concentrations and solution pH agrees with RO measurements of salt rejection and permeability. Our findings demonstrate that co-ion partitioning and membrane charge are important in determining salt transport through RO membranes.

4:00 pm (Room Palm BE, Wednesday)

Oral 155 - Forward Osmosis and Freeze Concentration for Mining and Metals Effluents

Noel Devaere (University of Toronto)*; Runlin Yuan (University of Toronto); Vladimiros Papangelakis (University of Toronto)

In Canada alone, water use in the mining and metals industry is over one billion cubic meters of intake and discharge per annum. Companies face increasing costs, tightening regulations, and mounting public pressure to reduce their water consumption, but there is a lack of low energy water recovery technologies for their high dissolved solids effluents.

Previously untested on these wastes, forward osmosis (FO) spontaneously extracts water from high concentration feeds with less irreversible fouling than reverse osmosis; however, draw solution regeneration remains a challenge. To solve this, we are developing a freeze concentration (FC) process: separation of an aqueous salt solution into ice and a reconcentrated draw solution. Compared to evaporation for high concentration draw solution reconstitution, freezing is more energy efficient and has the possibility to use environmental cooling to further reduce energy consumption. Thus, it is of interest to evaluate FO performance with draw concentrations where FC can operate.

Using simulation and experimental data, we demonstrate that forward osmosis with freeze concentration can recover water from waste samples with high osmotic pressures by employing a commercially available FO membrane. To achieve optimal FO operation, we showcase how water flux is increased while rejections are maintained by raising crossflow velocity and changing draw salts under osmotic gradient equivalent conditions.

4:30 pm (Room Palm BE, Wednesday)

Oral 156 - Predicting Efficiency And Elucidating Mechanisms For Organic And Inorganic Compound Removal By Reverse Osmosis And Nanofiltration Membranes Using Machine Learning

Nohyeong Jeong (Colorado State University)*; Tiezheng Tong (Colorado State University)

Predictive models for organic and inorganic compound removal by membrane separation are highly desirable for the design and selection of appropriate membranes. While machine learning (ML) models have been applied for such purposes, their reliability might be compromised by data leakage due to inappropriate data splitting. More importantly, whether ML models can truly understand the mechanisms of membrane separation has not been revealed. In this study, we evaluate the capability of the XGBoost model to predict organic and inorganic compound removal efficiencies of reverse osmosis and nanofiltration membranes. Our results demonstrate that data leakage leads to falsely high prediction accuracy. By utilizing a model interpretation method based on the cooperative game theory, we test the knowledge of XGBoost on the mechanisms of membrane separation via quantifying the contributions of input variables to the model predictions. We reveal that for organic compound removal, XGBoost possesses an adequate understanding of size exclusion, but its knowledge of electrostatic interactions and adsorption is limited. The XGBoost model also demonstrates that the transport of cations and anions is regulated by different ion and membrane properties, indicating that cations and anions might permeate through membranes independently. Our findings suggest that future work should focus more on avoiding data leakage, and that ML models have the potentials of not only predicting membrane performance but also deepening our understanding of mechanisms of solute transport during membrane separation.

Oral 27 – MEMBRANE FOULING II

2:00 pm (Room Palm CF, Wednesday)

Oral 157 - Fouling-Resistant Membranes with Tunable Pore Size Fabricated Using Cross-Linkable Copolymers with High Zwitterion Content

Samuel J Lounder (Tufts)*; Ayse Asatekin (Tufts); Luca Mazzaferro (Tufts University); Patrick Wright (Tufts University)

This work describes a new approach for synthesizing extremely fouling-resistant, zwitterionic membranes with controlled, tunable pore sizes that extend from ion separations (< 1 nm) to the ultrafiltration range (~2 nm). These membranes are manufactured by the UV treatment of high zwitterion content amphiphilic copolymers with cross-linkable functionality to stabilize the membrane selective layers, preventing excessive swelling and dissolution of copolymers containing as high as 80 wt% zwitterionic repeat units. Zwitterion weight fraction allows the tuning of membrane performance, with effective pore size and permeance both increasing with zwitterion content. The high zwitterion content membranes were remarkably fouling-resistant and demonstrated a salt-responsive behavior not previously observed with self-assembling zwitterionic copolymer membranes.

2:30 pm (Room Palm CF, Wednesday)

Oral 158 - Combined in-situ imaging and CFD to unravel the scaling mechanism in membrane distillation of high salinity brine

I-Min Hsieh (Texas Tech University); Bosong Lin (Texas Tech University); Hadis Matinpour (Texas Tech University); Mahdi Malmali (Texas Tech University)*

Membrane distillation (MD) has attracted a lot of attention for the desalination of hypersaline brine. To date, mineral scaling on the membrane surface is remaining as one of the significant challenges that limit its commercialization. Scaling might reduce the membrane throughput or lead to scaling-induced membrane wetting. In this presentation, I will present the design of a non-invasive apparatus to monitor the scaling phenomenon in vacuum MD (VMD) in real-time. This setup consists of optical imaging cameras on both feed and permeate sides of the membrane module, in-line turbidity, and conductivity meters to track the formation of the crystals and deposition on the membrane surface. This setup successfully helped us to 1) study the scaling mechanism both at high and low concentration polarization (CP) regions; 2) investigate the wetting phenomenon at the permeate side of the VMD; 3) evaluate the scaling mitigation by varying the flow hydrodynamics. The unique findings of this state-of-the-art experimental setup, integrated with computational fluid dynamics (CFD) hydrodynamic models provides unprecedented finding that can help to further shed light on the scaling deposition mechanism and crystal growth within the VMD channel for hypersaline brine desalination. The corresponding wetting phenomenon of different scalants, including calcium sulfate, calcium carbonate, and silica, have been studied and compared so the interplay between the membrane performance, scaling, and wetting can be further clarified. The real-time saturation index estimated by the ASPEN model of each scalant determined the onset of heterogeneous and homogeneous crystallization on the membrane surface, which agreed with our in-line monitoring. The SEM and XRD confirmed morphologies and crystallinity of the gypsum, calcite, and silica nanoparticles. The factors that influence scaling, like electrolytes' concentration, electrolyte-electrolyte interactions, and flow hydrodynamics were investigated.

3:00 pm (Room Palm CF, Wednesday)

Oral 159 - Fouling as an Opportunity: Exploiting the Properties of Ultrathin Membranes to Make Fouling-Based Sensors

James McGrath (University of Rochester)*

Ultrathin silicon membranes (50 nm - 400 nm thick) have many characteristics that distinguish them from conventional polymeric membranes. Among these are exceptional hydraulic permeabilities; well-defined, non-tortuous pores that are equal in size to the membrane thickness; and naturally small, 'chip-based', formats. With these properties, ultrathin membranes foul quickly in the presence of species large enough to occlude or constrict pores. While rapid clogging poses challenges for the application of ultrathin membranes in separations, it creates unique opportunities for the use of porous membranes as sensors. Here I will review advances on two membrane-based sensing concepts under development by our group: 1) the capture of extracellular vesicles (EVs) for 'digital' analysis of EV content and function, and 2) the rapid and specific capture of virus to cause a fouling-based hydraulic switch.

In the first example, we use ultrathin silicon nitride (100 nm thick) with a high density (~50 x 108 mm^2) of pores tuned to the sizes of diagnostically valuable small EVs (sEVs). Simple dead-end filtration captures and isolates single sEVs in pores allowing the simultaneous inspection of thousands sEVs by fluorescence light microscopy. Our work has revealed for the first time that individual sEVs have functional sodium-hydrogen exchangers that they use to maintain cargo at neutral pH as they transit biological microenvironments (Riazanski et al. (2022) Commun Biol 5, 13).

The second example is a membrane-based sensor that requires both viral size and affinity to produce a positive result. Proof-of-principle studies using Vaccinia demonstrate outstanding specificity (100%) and sensitivity (97%) (Klaczko, et al. (2022) Analyst, 147: 213-222). Our prototype device is powered by a pipette and requires no external reader for a positive indication suggesting the potential for use in a point-of-care assay to help manage infectious disease.

3:30 pm (Room Palm CF, Wednesday)

Oral 160 - Reactive Reverse Osmosis Membranes For Dissolved Silica Antifouling And Concentration Polarization Reduction By Catalytical Degradation Of Hydrogen Peroxide

Weiliang Bai (University of Texas at Austin)*; Ruizhe Xu (University of Texas at Austin); Adam Boehnke (University of Texas at Austin); Navid Saleh (University of Texas at Austin); Manish Kumar (University of Texas at Austin)

Reverse Osmosis (RO) filtration is an effective technology due to its high removal efficiency and mild working conditions. It is widely used in water treatment, especially in brackish and seawater desalination. However, the removal efficiency and permeate flux drops with time because of the foulant attached to the membrane surface. Hydrogen peroxide with a relatively low price has been employed as an oxidant in drinking water treatment with the AOP process. Previous research showed that membrane fabrication by metal oxide with the addition of hydrogen peroxide is efficient for antifouling when colloidal silica fouling, organic fouling, biofouling, or oil fouling appeared. Dissolved silica is presented in some brackish water sources, and they will deposit on the membrane surface because of concentration polarization. The deposition of silica decreases the permeate flow rate over time. There is no report on the effect of hydrogen peroxide catalytical decomposition on the influence of dissolved silica fouling.

In this project, we followed the previous result and produced RO membrane coating with polydopamine as well as the ex-situ produced copper oxide or manganese dioxide as the catalysts. The produced membrane showed a much-improved antifouling effect and lower concentration polarization modulus when hydrogen peroxide is presented in the feed water. Periodically added hydrogen peroxide could be catalytical decomposed by metal to generate oxygen microbubbles on the synthesized membrane. Oxygen bubbles take away the attaching foulant and increase the mixing within the feed channel to produce a lower concentration polarization modulus. The polydopamine layer helped protect membranes from damage by microbubbles and reactive radicals, which enabled long-term use in membrane filtration. This research provided a comprehensive evaluation for the application of copper and manganese oxides in the dissolved silica antifouling membrane on RO system.

4:00 pm (Room Palm CF, Wednesday)

Oral 161 - Scalable antifouling membranes via zwitterionic polymer brushes: examining the roles of brush thickness and density

Allyson L McGaughey (Princeton University)*; Sneha Srinivasan (Princeton University); Kofi SS Christie (Princeton University); Zhiyong Jason Ren (Princeton University); Rodney Priestley (Princeton University)

In virtually all membrane processes, fouling severely impacts performance – reducing permeability, increasing energy consumption, and increasing costs. Zwitterionic polymer brushes have shown great potential to prevent fouling with ultrathin, durable layers that preserve or improve permeability and rejection. However, conventional brush fabrication is complex and oxygen-sensitive, with limited potential for scale-up.

Recently, Cu^0-mediated atomic transfer radical polymerization (Cu^0-ATRP) has shown great potential as a facile, scalable fabrication method that can be used for a wide range of brush chemistries. Cu^0-ATRP can be done in air, under environmental conditions, for rapid brush growth. However, Cu^0-mediated ATRP has not previously been used for membrane modification.

We demonstrate the use of Cu^0-ATRP to form zwitterionic poly(sulfobetaine methacrylate) brushes on poly(vinylidene fluoride) (PVDF) layers and microfiltration membranes. Brush layers render spin-coated PVDF layers (contact angle (θ) = 85 ± 2°) hydrophilic (θ = 10 ± 1° after 1 h polymerization) and microporous PVDF membranes (θ = 131 ± 4°) highly wettable, absorbing droplets in ~10 s after just 10 min polymerization and ~2 s after 1 h. We systematically investigate the effect of plasma treatment and polymerization time on material properties including hydrophilicity, brush thickness, brush density, and permeability.

While previous studies generally varied only brush thickness and examined only one foulant type, we evaluate fouling resistance to multiple foulant types (oils, organics, proteins, biofouling) as a function of pH and salinity. We characterize foulant-membrane interactions via colloidal probe force microscopy and relate results to dynamic fouling in microfiltration. The results of this work provide fundamental understanding of the effects of material properties on performance for a range of applications, towards resource and water recovery from challenging waste streams.

4:30 pm (Room Palm CF, Wednesday)

Oral 162 - Raman Chemical Fingerprints for Real-time Detection of Membrane Fouling

Danielle Park (University of Colorado Boulder)*; Omkar D. Supekar (Department of Mechanical Engineering, Department of Electrical, Computer, and Energy Engineering; University of Colorado Boulder); Alan R. Greenberg (Membrane Science, Engineering, and Technology (MAST) Center, Department of Mechanical Engineering,

University of Colorado Boulder); Juliet T. Gopinath (Department of Electrical, Computer, and Energy Engineering, Department of Physics; University of Colorado Boulder); Victor M. Bright (Department of Mechanical Engineering, University of Colorado Boulder)

Despite an increased focus on alternative approaches, the dominant technology for desalination remains reverse osmosis (RO). Among the challenges in developing more efficient membrane-based RO desalination is fouling, which results in degraded performance, increased energy consumption and higher operating costs. Reliable real-time detection of early-stage membrane fouling can provide an important, alternate means to address these concerns since fouling remediation strategies can be most effectively employed before fouling becomes extensive.

In recent work, we demonstrated real-time detection of inorganic fouling with chemical identification using Raman spectroscopy. The noninvasive methodology based on a light-matter interaction can provide a precise chemical fingerprint of the fouling species with sub-micron spatial resolution and a rapid temporal response. In previously published work, successful results for scaling detection and monitoring of membrane cleaning were obtained under realistic RO operating conditions; however, the effects of systematic changes in feed concentration and composition were not considered in these initial experiments.

In this study we utilize new custom designed and more versatile Raman microscope-RO flow cell instrumentation. The improved capabilities enable expansion of the previous work to consider fouling detection of multi-component feeds at low concentrations over extended time scales. In addition, the ability to modify the x-y detection coordinates allows for comparison of fouling kinetics at different locations along the flow path. Overall, results obtained using this Raman-based sensor methodology show considerable promise for enhanced performance characterization and new insights regarding fouling initiation and mechanisms.

Oral 28 – CONTAMINANT REMOVAL FROM WATER

2:00 pm (Room Palm AD, Wednesday)

Oral 163 - Development Of Highly Selective Biocatalytic Membrane Reactors For The Degradation Of Pesticides

Lidietta Giorno (National Research Council Of Italy - Institute On Membrane Technology (CNR-ITM))*; Giuseppe Vitola (National Research Council Of Italy, Insitute On Membrane Technology, CNR-ITM); Rosalinda Mazzei (National Research Council Of Italy, Insitute On Membrane Technology, CNR-ITM)

The use of organophosphate pesticides in agriculture has led to the accumulation of these substances in the environment. Pesticides belong to the category of recalcitrant microcontaminants. Chronic exposure to such components can cause serious health problems. Clean, highly selective, and efficient methods for pesticide degradation are needed. Enzymes such as phosphotriesterase from extremophiles have demonstrated these properties. However, enzymes are labile macromolecules and are rapidly deactivated. Immobilization of enzymes in membranes has been shown to increase the catalytic stability of enzymes. Therefore, the immobilization of phosphotriesterase on various membranes was systematically studied. The enzyme was immobilized on both hydrophilic and hydrophobic membranes by covalent bonding. Hydrophilic polymer membranes were found to be the most suitable for maintaining enzyme activity and stability. A regenerated cellulose membrane was selected for the development of a biocatalytic membrane reactor. The enzyme is immobilized in the pores of flat-sheet membrane. The paraoxon pesticide solution (up to a concentration of 1 mM) was passed through the enzyme-loaded membrane. Each membrane pore worked as a continuous microreactor and the biodegraded solution was collected in the permeate side. With optimal biocatalyst amount and residence time, complete conversion of the pesticide can be achieved.

The biocatalytic membrane reactor was also tested with real streams consisting of surface water or olive mill vegetative waters spiked with a known amount of pesticide. The half-life time of the biocatalytic membrane used with surface water and olive mill water were 69 and 22 days, respectively. These values were much higher than those of the free enzyme, whose $t_{1/2}$ in surface water and olive mill water were 22 and 4.5 days, respectively.

2:30 pm (Room Palm AD, Wednesday)

Oral 164 - Negative Rejection in Polyelectrolyte Multilayer Nanofiltration Membranes for Selective Ion Recovery

Brielle Januszewski (Yale University)*

Technologies that separate ions of interest with high precision can effectively recover high-value ions in brines, thus contributing to a circular economy and preventing the release of toxic ions into the environment. Membranes can operate continuously and require few chemical inputs, offering an alternative to resource-intensive ion separation methods such as precipitation or ion exchange. Here, we demonstrate a commercial polyamide nanofiltration (NF) membrane modified with two polyelectrolytes of opposite charges, poly(allylamine hydrochloride) (PAH) and poly(styrene sulfonate) (PSS), to extract monovalent species from multi-component feed streams. Our results establish that polyelectrolyte multilayers impart charge to the membrane that leads to high positive rejection of divalent cations (> 95%) and highly negative rejection of monovalent cations, providing high selectivity between monovalent and divalent cations. Adjusting the pH of the feed water towards the pK_a (~9) of the PAH amine groups renders the membrane less selective between monovalent and divalent cations. Moreover, when increasing the ratio between divalent and monovalent cations in the feed solution, we notice an improvement in selectivity and negative rejection, likely caused by an increase in membrane charge due to accumulation of magnesium ions near the surface. Finally, we apply these membranes for lithium recovery, with initial experiments reporting a Li^+/Mg^{2+} selectivity of 39. Overall, we show charge-based exclusion mechanisms are most influential on membrane selectivity and that these membranes can extract monovalent ions even when present at very low concentrations. This work contributes to the fundamental understanding of mechanisms of ion separations and can inform future membrane design that seeks to separate ions and purify feed streams.

3:00 pm (Room Palm AD, Wednesday)

Oral 165 - MXene-functionalized Thin-film Composite Membranes for PFAS Removal from Water

Milad Rabbani Esfahani (The University of Alabama); Tin Le (The University of Alabama)*

The historical widespread industrial use of per- and polyfluoroakyl substances (PFAS) has resulted in contamination of many water sources around the United States. Due to their negative health and developmental effects, PFAS removal is an emerging research area in water purification. We demonstrate the functionalization of thin-film composite (TFC) hollow fiber nanofiltration (HFN) membranes by MXene-nanosheets during the interfacial polymerization (IP) process for enhanced removal of Perfluorooctane sulfonic acid (PFOS) from water. A MXene-polyamide selective layer was fabricated on top of polysulfone hollow fiber support via IP of trimesoyl chloride (TMC) and a mixture of piperazine (PIP) and MXene nanosheets. Incorporating MXene nanosheets during the IP process tuned the morphology and negative surface charge of the selective layer, resulting in an increase in PFOS rejection from 72% (bare TFC) to more than 96% (0.025% (wt.)-MXene-TFC) while the water permeability was also increased from 13.19 (bare TFC) to 29.26 LMH/bar (0.025% (wt.)-MXene-TFC). Our results demonstrate that both electrostatic interaction and size exclusion are the two main factors governing the PFOS rejection and both are determined by PA selective layer structural and chemical properties. MXene nanosheets showed very promising capability as a 2-D additive for tuning the structural and chemical properties of PA layer at the permeability-rejection tradeoff.

3:30 pm (Room Palm AD, Wednesday)

Oral 166 - Graft Polymerization Using Oppositely Charged Monomers On NF Membrane For The Efficient Removal Of Carbamazepine From Wastewater Effluents

Brhanu Kelali Desta (Ben-Gurion University of the Negev)*; Akshay Modi (Ben Gurion University of the Negev); Ron Kasher (Ben Gurion University of the Negev)

Carbamazepine (CBZ) is a pollutant that is discharged to the environment from the pharmaceutical industry, agriculture, and commercial items. It is highly persistent in the environment and detected in crops irrigated with reclaimed wastewater. Poor rejection of CBZ by commercial NF membranes may pose limitations on their practical applications. In this study, we showed improved CBZ rejection efficiency of a commercial nanofiltration membrane without compensating the membrane's pure water permeance. Different membranes were modified by redox-initiated graft polymerization using oppositely charged methacrylate monomers with varied ratio of the negatively charged methacrylic acid (MA) and positively charged [2-(methacryloyloxy)ethyl]-trimethylammonium (MOETMA) monomers. Fourier-transform infrared spectroscopy analysis confirmed the membrane modifications with the appearance of characteristics IR peaks. The graft polymerization endowed the modified membranes with excellent hydrophilicity, and 15% higher pure water permeance than the pristine membrane which is explained by their surface hydrophilicity. The CBZ rejection efficiency of the membranes was measured using a solution that simulated secondary-treated municipal wastewater effluents spiked with 0.2 ppm CBZ. The membrane grafted with a high proportion of MOETMA over MA (9:1) showed a significantly high CBZ rejection efficiency of 87% compared to the rejection efficiency measured for the pristine membrane (55%). These results indicate that the controlled redox-initiated graft polymerization method described here improves CBZ removal efficiency with concurrent improvement in the membrane pure water permeance, which has high implications for advanced treatment of reclaimed wastewater providing safer and cleaner water for irrigation.

4:00 pm (Room Palm AD, Wednesday)

Oral 167 - Modeling of RO System Water Treatment Operation for Nitrate and Salt Removal using Long-Short Term Memory (LSTM) Machine Learning Model with Attention Coefficient

Yoram Cohen (University of California, Los Angeles)*; Muhammad Bilal (University of California, Los Angeles); Yang Zhou (Shanghai University)

The performance of an advanced autonomous RO water treatment system for wellhead water treatment for salinity and nitrate removal was investigated via a machine learning (ML). The water treatment system, deployed in Salinas Valley (CA) had a production capacity of ~ 2500 gallons/day, treating groundwater of salinity and nitrate levels of ~1500 mg/L TDS and ~55 mg/L nitrate as N, respectively. The RO system operated intermittently with partial recycle. System performance was analyzed with the aim toward forecasting permeate nitrate and salinity levels via a forecasting machine learning (ML) approach accompanied by a comprehensive exploratory analysis and Long-Short Term Memory (LSTM) modeling approach with and without an attention coefficient. Data-driven LSTM models were developed for the permeate nitrate and salinity levels for the different modes of system operation (i.e., startup, permeate production, flushing, and shutdown) with twenty attributes from a multitude of RO system sensors (e.g., pressure, flow rate, conductivity, nitrate) in different system locations. LSTM model showed excellent performance for each operational mode (i.e., R^2 value of equal or greater than 0.92). In addition, a minimal set of model attributes was identified and ranked via Pearson's correlation coefficient and feed forward feature selection (FFFS) approaches. Subsequently, unsupervised self-organizing maps (SOM) approach was applied to discover similarities among model attributes. The machine learning modeling workflow and model developed in this study demonstrated a multitude of benefits including: (a) determining the deviation of nitrate from the baseline regulatory value, (b) forecasting performance of an RO system, under field conditions, with respect to nitrate permeate and salt passage, (c) assessing the correlation between nitrate permeate concentration and salt passage, and (d) providing the basis for model-based control for the RO system.

4:30 pm (Room Palm AD, Wednesday)

Oral 168 - Nanofiltration and temperature swing reverses osmosis for the removal of herbicide 2-methyl-4-chlorophenoxyacetic acid (MCPA) from saline industrial wastewater

Mikel Duke (Victoria University, Melbourne)*; Greg Weston (Tri-Tech Chemical Co. Pty Ltd, Australia); Stephen Gray (Victoria University); Xing Yang (Victoria University); Jianhua Zhang (Victoria University)

Herbicides are widely utilised for crop protection in agriculture, but have resulted in the presence of their residues in various environments [1]. One of the popular types are chlorophenoxy herbicides such as 2-methyl-4-chlorophenoxyacetic acid (MCPA) used for the control of broad-leaved weeds. Since 1970s, the potential for chlorophenoxy herbicides to cause certain forms of cancer in humans has come under increasing scrutiny [2]. Attempts have also been made to treat wastewaters containing MCPA. The majority of herbicide compounds have molecular weights greater than 200 Da and sizes in the range of ions (close to 1 nm) so reverse osmosis (RO) and nanofiltration (NF) are the most appropriate pressure-driven membrane processes but have not been considered. Conventional reverse osmosis (RO), a novel temperature-swing RO (TSRO), and tailored nanofiltration (NF), were investigated to target the removal of MCPA from HCl-rich wastewater. Batch concentration by RO at 17.5 bar using BW30 and SW30 membranes showed similar performance, where BW30 showed rejections of 95.3% for MCPA and 25.9% for acid, but initial flux of 13 $L.m^{-2}.h^{-1}$ declined by 82% upon reaching 80% volume recovery. For the novel TSRO using BW30 and swing temperature of 40-45°C based on MCPA solubility tests, the significant flux decline was avoided while showing similar MCPA rejection, but with slightly higher acid rejection of 37%. Pilot scale NF using NF270 and NF90 membranes showed decreased rejections of both MCPA and acid. The NF270 showed rejections of 71.3% for MCPA and 16.8% for acid. But despite lower MCPA rejection than RO (and NF90), was sufficiently high to potentially meet MCPA discharge requirements at high flux to 80% volume reduction. RO and NF are therefore suitable for MCPA removal from saline (HCl-rich) wastewaters, but novel temperature swing RO process offers potential to achieve higher MCPA rejection simultaneously to higher overall flux at practical water recoveries.

Oral 29 – WATER REUSE

2:00 pm (Room Dolores, Wednesday)

Oral 169 - Electroconductive Filters and Membranes: Future for Desalination and Wastewater Treatment

Swatantra Pratap Singh (IIT BOMBAY)*; Najmul Barbhuiya (IIT Bombay); Utkarsh Misra (IIT Bombay)

Fouling and selectivity permeability tradeoff are the leading challenges in the membrane-based desalination and wastewater treatment technology. Laser-induced graphene (LIG) is a three-dimensional porous electrically conductive graphene material printed on the polymer substrates with a 10.6 mm CO-2 laser in a single step and chemical free process. Polyethersulfone(PES) is a highly used polymer in the membrane technology for manufacturing microfiltration, ultrafiltration and reverse osmosis membranes. A single-step conversion of doped PES membranes to LIG filters and further UF membrane has been demonstrated with excellent catalytic and anti-biofouling properties. These electrically conductive filters and membranes achieved 6-log removal of mix culture of bacteria with applied voltage along with the generation of H_2O_2. Furthermore, these surfaces are shown better antibiofouling properties. These conductive membranes can be tuned for the in-situ selective removal of the contaminants. The "catalytic mechanisms of these doped LIG filters will be explained, including the 3D texture effect of LIG. These filters and membranes (single-step laser printed) have great potential in desalination and wastewater treatment.

2:30 pm (Room Dolores, Wednesday)

Oral 170 - Chlorination And Autopsy Of Fouled Engineering-Scale Reverse Osmosis Membranes

Bianca M. Souza Chaves (The University of Arizona)*; Mohammed Alhussaini (The university of Arizona); Varinia V Felix (The University of Arizona); Andrea Achilli (The University of Arizona)

The disposal of end-of-life (EoL) reverse osmosis (RO) membranes is still a critical challenge and chlorination has emerged as an alternative to extending the lifespan of RO membranes for water reuse applications. Engineering-scale results have shown that controlled chlorine exposure is effective in removing irreversible fouling of EoL RO membranes and in restoring membrane permeability without compromising selectivity. However, the exposure dose is not conservative and changing the concentration or duration for the same exposure dose could provide different results and/or considerably damage the polyamide layer. To gain a comprehensive understanding of how chlorination changes the EoL RO membrane characteristics and mass transfer properties, four EoL RO spiral-wound elements were removed from an engineering-scale system, unrolled, cut in coupons, and tested in a flow-through cross-flow configuration bench-scale system with reclaimed wastewater effluent. The goal is to elucidate the membrane oxidation pathway and the mechanisms of organics and inorganics permeation in chlorinated EoL RO membranes. To accomplish that, the fouled coupons were oxidized by progressive exposure to free-chlorine doses (up to 10,000 ppm-h). The apparent water, conductivity, ion, and organic permeability coefficients and rejection were calculated and related to the chlorine dose. Size-exclusion chromatography provided information on the molecular weight cut-off, and EEM fluorescence on the type of organics permeating through the chlorinated membrane. SEM images and XPS spectroscopy were used to determine the elemental composition and morphology of the modified membrane. Bench-scale results were also compared with engineering-scale chlorinated membranes to investigate how the oxidation compromised the RO membrane integrity and selectivity. Results on the apparent water, salt, and organic permeability, rejection, and changes in membrane characteristics will be discussed in the presentation.

3:00 pm (Room Dolores, Wednesday)

Oral 171 - A balanced-charged polyelectrolyte complex nanofiltration membrane: mitigating scaling in effluents desalination

Hao Huang (Ben Gurion University of the Negev)*; Roy Bernstein (Ben Gurion University)

High salinity and high sodium adsorption ratio (SAR) limit wastewater effluent application for direct irrigation. Effluent desalination using nanofiltration (NF) is an attractive approach to overcome these challenges and produce high-quality effluent for unrestricted irrigation. However, the selectivity of the typical polyamide NF membranes is mostly toward divalent ions. Consequently, they produce effluent with high SAR levels. Moreover, due to the high phosphate ions selectivity, $CaPO_4$ scaling limits the recovery ratio of NF during effluent desalination.

Our research aims to develop a polyelectrolyte complex NF membrane with moderate removal of all ions (50-70%) to achieve high recovery desalination and produce effluent with a composition sustainable for irrigation.

The membranes are fabricated by coating an ultrafiltration membrane with a homogeneous layer of polysodium 4-styrene sulfonate (PSS) and polyethyleneimine (PEI). The ions selectivity is tuned by adjusting the weight ratio between PSS and PEI and the permeability by changing polymer concentration and applying pre-treatment steps.

A balanced surface charged membrane obtained at PSS:PEI ratio of 1:1.3 resulted in symmetric rejection for divalent anions and cation (51% for $MgCl_2$, 62% for Na_2SO_4, 53% for $MgSO_4$, and 41% for NaCl). In addition, chemical crosslinking was applied to increase the long-term stability of the NF membrane. Moreover, the initial permeability of the membrane was 1.3 LMH/Bar and was increased to 7.1 LMH/Bar by optimizing the polymer concentration and soaking the membrane in glycerol solution before the membrane fabrication to eliminate pore-collapse. The selectivity of the NF membrane for mixed ion solutions is currently investigated. Ultimately, the membrane will be studied for AnMBR effluent desalination.

3:30 pm (Room Dolores, Wednesday)

Oral 172 - Degradation of Polyamide Thin Film by Chlorine and Peracetic acid: A Combined QCM-D and AFM Study

Tashfia M Mohona (University At Buffalo)*; Ning Dai (University At Buffalo); Prathima Nalam (University At Buffalo)

Membrane fouling remains a critical challenge in maintaining the high and reliable performance of membrane processes. Chlorine is a traditional disinfectant effective in suppressing or removing biofilms, but it can degrade the active polyamide layer in thin-film composite nanofiltration and reverse osmosis membranes. In this study, we evaluated the compatibility of a new disinfectant, peracetic acid (PAA), with polyamide thin films using molecular-level techniques, and compared it with that of chlorine. Polyamide degradation kinetics was examined by quartz crystal microbalance with dissipation (QCM-D) and atomic force microscopy (AFM). Polyamide films were first synthesized with controlled thickness using a molecular layer-by-layer deposition approach. QCM-D experiments revealed that polyamide degradation by both oxidants featured distinct sorption and degradation phases. Chlorine (100 ppm) exposure led to a short (1.4 – 3.5 min) sorption phase following Boltzmann-sigmoidal behavior while PAA (1000 ppm) exposure led to a two-stage sorption spanning between 3 – 33 h. In the degradation phase, chlorine-exposed films showed two-stage exponential decay, reaching degradation equilibrium at ~5 hours after losing ~50% of the total initial mass. In contrast, PAA-exposed films followed a Boltzmann sigmoidal degradation behavior, losing only about ~20% of the total initial mass after ~10 h. AFM imaging in stagnant oxidation solutions showed that chlorine exposure (10 h) resulted in progressive loss in polyamide integrity via the formation of pits and cracks, while PAA-exposed films showed no detectable change in morphology even after 148 h. Overall, our results support the much greater compatibility of polyamide with PAA than with chlorine. Moreover, the molecular length-scale characterization revealed PAA-induced small changes to polyamide that should be further investigated to fully understand the implication of PAA disinfection on the long-term stability of membranes.

4:00 pm (Room Dolores, Wednesday)

Oral 173 - Solute-solute selectivity in membrane separation: definition, measurement, and process scale evaluation

Ruoyu Wang (vanderbilt university)*; Shihong Lin (Vanderbilt University)

Precise solute-solute separation using membranes has recently emerged to become one of the frontiers for membrane-based separation due to the increasing demand for selective separation in resource recovery and extraction from aqueous feed streams. Example applications of selective separation include the extraction of lithium from brine, recovery of nutrients from wastewater, and recycling of acid or base from process streams. Most existing studies focused on fabricating novel membranes for selective solute-solute separation, while limited recent studies have also investigated the fundamental mechanisms to enhance solute-solute selectivity. Nonetheless, certain basic but critical questions regarding the measurement or even just the definition of solute-solute selectivity remain unsettled. For instance, it is unclear if solute-solute selectivity can be compared when obtained using different experimental methods and conditions. It is also uncertain which of the existing definitions of selectivity is more practically consistent with treatment goals in respective applications. This presentation aims to address these questions and provide new insights to establishing the theoretical framework of quantifying solute-solute selectivity in membrane-based precise separation. Additionally, we will also discuss how to evaluate the process scale solute-solute selectivity (Li/Mg separation as an example) based on data measured using small membrane coupons.

4:30 pm (Room Dolores, Wednesday)

Oral 174 - Effect of Pore Defects on Membrane Rejection in Pilot-scale Membrane Distillation

Mukta Hardikar (The University of Arizona)*; Luisa Ikner (University of Arizona); Varinia V Felix (The University of Arizona); Luke Presson (University of Arizona); Andrew Rabe (University of Arizona); Kerri Hickenbottom (U. Arizona); Andrea Achilli (The University of Arizona)

Membrane distillation (MD) is a potentially suitable process for treating reverse osmosis concentrate to produce distillate and reduce volume of concentrate to be disposed. High rejection of contaminants is an essential parameter for potable reuse of water treated in MD. Theoretically, MD can achieve 100% rejection of non-volatile components such as solutes and pathogens because only the vapor phase permeates through the membrane. Rejection of organics and salts has been reported in a wide range (90-99.99%) but its resolution is limited by the solubility of the contaminant in the feed and the detection limit of contaminants in the distillate. Also, thus far, pathogen rejection in pilot-scale MD has not been investigated. At the pilot-scale, the larger membrane area increases probability of transport of contaminants through pore defects while it decreases the water flux and dilution of contaminants in the distillate. Studying rejection of pathogens also overcomes the solubility limit of the contaminant in the feed. In this work, membrane rejection of non-volatile components in a pilot-scale air-gap MD (AGMD) element is quantified using bacteriophages. Reclaimed wastewater was used as feed and was spiked with two bacteriophages MS2 and PhiX174 at approximately 4×10^5 PFU/mL concentration. Membrane rejection of bacteriophages, cations and organics was quantified at a constant flowrate and temperature difference between the feed and distillate (40/20 °C) and low and high vacuum in the air gap. Membrane rejection of bacteriophages was observed to be 0.5-log10 higher than cation rejection. Rejection of organics, inorganics and bacteriophages reduced with increase in vacuum in the air gap, which suggests presence of pore flow through pore defects. Interestingly, exposure to high feed temperature (70 °C) did not lead to an instantaneous decrease in feed virus concentration suggesting that longer operating times are required to decrease virus concentration in the feed.

Oral 30 – MODULE MODELING AND DESIGN

2:00 pm (Room Xavier, Wednesday)

Oral 175 - Computational Fluid Dynamics (CFD) Modeling of 3D-Printed Spacers and Patterned Membranes in Reverse Osmosis and Nanofiltration Modules

David Ladner (Clemson)*; Zuo Zhou (Clemson University)

3D printing has been under development for creating spacers in reverse osmosis (RO) and nanofiltration (NF) modules. Patterning technology has also been under investigation for creating novel membrane topologies. These techniques may be advantageous because they can handle complicated spacer geometries and membrane topologies and simplify the manufacturing process. One manufacturing method uses 3D printing to directly attach spacers to the membrane surface. The novelty of this method is that thin spacers can be built, resulting in thin flow channels between membrane leaves. Within the same cylindrical module volume, more layers can be packed and more membrane surface area can be added to achieve higher permeate flow rates. The goal of this study is to design efficient models that help us discover the best parameters for 3D-printed spacers and patterned membranes. Spacer heights and spacer patterns are directly related to longitudinal pressure drop and concentration polarization (CP). There is typically an inverse correlation between pressure drop and CP because energy is required to create mixing. The goal, then, is to find geometries where mixing is induced and CP is reduced under low pressure-drop regimes. Computational fluid dynamics (CFD) is useful in helping expedite the process of discovering the best designs. Simulations investigated spacer shapes including regular cylinders, elliptical cylinders, and airfoils. Ellipses with a length:width ratio of 2.4 while maintaining a between-feature distance of 6 mm were optimal for minimizing pressure drop. Spacer heights ranging from 200 μm to 500 μm were simulated to discover the height to achieve the same pressure drop as a 30 mil conventional spacer. Simulation results indicate that 3D-printed spacers with elliptical design can greatly increase water productivity in a spiral-wound module by increasing packing capacity. These designs also reduce CP by 21% by improving the hydrodynamics, compared to empty channels.

2:30 pm (Room Xavier, Wednesday)

Oral 176 - Effect Of Packing Nonuniformity On The Performance Of Hollow Fiber Membrane Gas Separation Modules Fabricated From Fiber Tows

Lili Sun (University of Toledo); Grigorios Panagakos (Carnegie Mellon University); Glenn Lipscomb (University of Toledo, Toledo)*

Membrane processes are the preferred option for many gas separations. Hollow fiber membranes commonly are used in these processes. Commercial hollow fiber membrane modules often are fabricated from small groups of fibers (i.e., a tow) that are arranged into a larger cylindrical bundle.

The use of fiber tows can lead to packing non-uniformity in two ways: 1) the spacing between tows and the surrounding case and 2) the spacing between tows within the bundle. This nonuniformity can result in nonideal fluid distribution in the shell and be detrimental to module performance.

Full three dimensional models of flow and mass transfer within fiber bundles formed from tows have been developed. These models account for the detailed fiber arrangement within modules and are solved to evaluate gas separation module performance in terms of stage cut and required membrane area as a function product gas purity for carbon capture and nitrogen production applications.

Results are presented illustrating the effects of varying fiber packing within a tow and the packing of tows. The results suggest smaller tows are preferred over larger tows, at fixed overall fiber packing, and the inter-tow packing should be comparable to the intra-tow fiber packing.

To reduce the computational cost of simulating large fiber bundles, the use of an equivalent planar bundle is proposed. A simple procedure for determining the geometry of the equivalent planar bundle is described and validated through comparisons of calculated performance metrics with results obtained from full three dimensional simulations.

3:00 pm (Room Xavier, Wednesday)

Oral 177 - Examining The Impact Of Acid Strength On Process Efficiency And Product Usability In Ammonia Recovery From Wastewater Using Membrane Contactors

Abhishek Dutta (University of British Columbia)*; Jongho Lee (University of British Columbia)

Dissolved ammonia in wastewater has been touted as a potential source to supplement commercial ammonia. Membrane contactors (MCs) that employ an acid draw solution (e.g., H_2SO_4) offer spontaneous and selective ammonia recovery from diverse ammonia-rich wastewaters (feed) in the form of ammonium fertilizers [e.g.,$(NH_4)_2SO_4$]. However, most MC processes employ excess acid to ensure high ammonia recovery, yielding a highly acidic product that is inappropriate to be applied as fertilizer and may damage the crops. In our study, we explore the impact of draw acid strength on product quality as fertilizer and ammonia recovery and identify suitable acid strengths to maximize both parameters. With a hydrophobic PVDF membrane placed between ammonia-rich aqueous solution (feed) and sulfuric acid (draw), we showed >95% ammonia recovery using a high strength acid (i.e., excess acid) under isothermal conditions. The nearly-invariant ammonia fluxes for different flow rates of draw solution confirmed the immediate acid-base reaction at the membrane-draw interface. When a low-strength acid was employed, on the other hand, the product of nearly neutral pH was obtained. However, a considerable reduction in the ammonia flux and consequently the ammonia recovery was observed, indicating the emergence of additional ammonia transfer resistance in the draw. We then developed an enhanced mass transfer model that describes the transfer resistance of ammonia in the draw solution arising from an incomplete acid-base reaction at the membrane-draw interface. After validating the model, we extended our model to show that (1) a trade-off exists between draw acid strength and ammonia flux, (2) optimal draw acid concentration can be identified to achieve usable pH of the ammonium salt as fertilizer without compromising its yield. Our study provides a guide for optimal acid strength selection for diverse wastewaters, reducing the downstream processing cost for ammonia fertilizer produced by MCs.

3:30 pm (Room Xavier, Wednesday)

Oral 178 - Accelerating The Optimization Of Turbulence-Promoting Spacers Using Computational Fluid Dynamics, Surrogate Modeling, And Machine Learning

Zachary Binger (The University of Arizona)*; Andrea Achilli (University of Arizona)

Turbulence-promoting spacers are an integral component of most membrane systems for providing both structural integrity and increased mixing for maximizing water recovery. To guide the development of future spacer shapes to maximize mass transfer and minimize pressure losses, it is imperative to accurately quantify the impact of such spacer designs. This can be done computationally, however, existing modeling approaches like computational fluid dynamics (CFD) can be computationally intensive and alone are impractical for wider-scoped surveys like optimization investigations. In this work, a multi-scale modeling framework for bridging the gap between high resolution but computationally intensive CFD simulations and more rapidly deployable models for predicting water recovery and energy consumption of membrane processes is demonstrated. Data collected from CFD simulations is used to train neural networks to replace existing physics-based or empirically derived models used in more lightweight modeling approaches to improve the accuracy, speed, and flexibility these models. This data-driven, bottom-up surrogate modeling approach can be used to capture the accuracy of higher resolution CFD modeling and translate it to more computationally lightweight software to perform fundamental studies into the hydrodynamics and mass transport in membrane modules. These surrogate models can then also be paired with optimization algorithms to find optimal spacer shapes that both maximize mass transfer and minimize pressure losses in membrane channels. This framework was used to accelerate the creation and testing of over 200 unique spacer shapes by varying six geometric parameters and simulating these designs over a range of velocities and water compositions. Overall, this research aims to improve the modeling of membrane systems that utilize turbulence-promoting spacers but also demonstrate exciting avenues to broaden the application of machine learning in water treatment research.

4:00 pm (Room Xavier, Wednesday)

Oral 179 - Flow Dynamics In A Spacer-Filled Membrane Distillation Channel

Bosong Lin (Texas Tech University); Hadis Matinpour (Texas Tech University); Mahdi Malmali (Texas Tech University)*

The role of the net-like spacer in a membrane module is to generate local turbulence and mixing in the feed channel and to mitigate polarization and fouling. It is generally expected that the addition of the spacer results in improvement in membrane performance. In the case of membrane distillation (MD), the induced mixing enhances mass and heat transfer and reduces the temperature and concentration polarization. Including the spacer in the feed channel can also pose the membrane to some risks. The wrong choice of spacer geometry can make the fouling and scaling more severe as the spacer can create more stagnant regions at the contact points of the spacer and membrane. One strategy for fouling and scaling mitigation is to optimize the spacer geometry that can generate a proper hydrodynamic design. The key parameter to such hydrodynamics is the onset of turbulence and vortices near the membrane.

This work aims at studying hydrodynamics in a spacer-filled channel, with an emphasis on the characterization of local turbulence. The 2-D flow velocity pattern in a spacer-filled channel is examined using micro Particle Image Velocimetry (PIV). The high-resolution flow measurement reveals the alterations in the flow direction and pattern due to the presence of the spacer. The velocity profile at each measurement plane suggests that the flow is parallel to the filaments. Moreover, turbulence kinetic energy and vortices derived from 2-D PIV imaging indicate the local mixing takes place in the middle of the channel. The velocity in the z-direction obtained using the continuity equation also suggests that the mixing of flow near and away from the membrane is not sufficient. A similar mixing behavior is confirmed in a 3-D computational fluid dynamics (CFD) simulation. The result suggests that to improve hydrodynamics, it is necessary to enhance the mixing near the membrane and bring the vortices closer to the wall.

4:30 pm (Room Xavier, Wednesday)

Oral 180 - Organic Matter Removal For Reverse Osmosis Fouling Mitigation: From Lab Prototype And Numerical Simulation To Pilot Plant

Yunqiao Ma (Nanyang Technological University)*; Tao Hua (Nanyang Technological University); Thien An Trinh (Nanyang Technological University); Rong Wang (Nanyang Technological University); Jia Wei Chew (Nanyang)

This project uses heated aluminum oxide particles (HAOPs) as a novel adsorbent for organic foulant removal to mitigate reverse osmosis (RO) membrane fouling. Since its launching in 2018, experimental and numerical simulation works have been conducted which enabled the subsequent scale-up to the pilot plant.

Experiment-wise, a prototype microgranular adsorptive filtration (μGAF) system was constructed employing a 7-bore ceramic membrane as the primary membrane and either HAOPs or powdered activated carbon (PAC) as the pre-deposited dynamic membrane. Pre-treat was done for membrane bioreactor (MBR) effluent from a full-scale water reclamation plant. The downstream RO performance and membrane fouling potential of the treated effluent were assessed. Results indicated that HAOPs were more effective in removing biopolymers (like polysaccharides and proteins) and phosphorus/fluoride, and virtually eliminated fouling of the primary membrane. Moreover, HAOPs-treated effluent resulted in a 43% decline in RO permeate water flux over 5 d of continuous filtration, as opposed to 62% flux decline for untreated or PAC-treated effluent. These results demonstrate the effectiveness of the HAOPs-based μGAF process as a pre-treatment for improving RO recovery.

As a supplement to experiments, molecular dynamics simulation was conducted to understand the competitive adsorption of constituents of effluent organic matter (EfOM) on HAOPs. Quantitative analysis reveals that most EfOM constituents anchor onto HAOPs despite steric hindrance due to high concentration. Adsorbed foulants exhibit much lower mobility and flexibility, indicating excellent adsorption capability of HAOPs before becoming fully saturated.

The pilot plant has been commissioned in 2021, and optimization of operating conditions is ongoing. It locates at Ulu Pandan Reclamation Plant (under PUB, Singapore's National Water Agency).

DOI: https://doi.org/10.1016/j.desal.2021.115016; https://doi.org/10.1016/j.memsci.2019.117651

NAMS 2022 - SPONSORS

NAMS 2022 - SPONSORS

DIAMOND LEVEL

National Science Foundation

University of Texas at Austin

GOLD LEVEL

ExxonMobil

Membrane Science, Engineering & Technology (MAST) Center

MilliporeSigma

Prometheus Fuels

SILVER LEVEL

Air Products

Chevron

DOW

Honeywell - Universal Oil Products

Membrane Technology & Research

Oak Ridge Associated Universities

The University of Arizona

BRONZE LEVEL

Ameridia Innovative Solutions Inc.

Gale & DB Battacharyya

Hiden Isochema

NL Chemical Technology Inc.

MDPI AG (Membranes Journals)

Evonik Chemicals

Georgia Institute of Technology

W. L. Gore & Associates

Sterlitech

EXHIBITORS

Anton Paar

Porometer

Surface Measurement Systems

Membrane Science, Engineering & Technology (MAST) Center

The University of Texas at Austin
Cockrell School of Engineering

The University of Texas at Austin
McKetta Department of Chemical Engineering
Cockrell School of Engineering

The University of Texas at Austin
Civil, Architectural and Environmental Engineering
Cockrell School of Engineering

ExxonMobil

Energy Lives Here™

ExxonMobil is a proud sponsor of NAMS 2022

https://energyfactor.exxonmobil.com/

NSF I/UCRC • Membrane Science, Engineering & Technology Center

MAST

A global leader in membrane research for over 30 years.

The Membrane Science, Engineering and Technology (MAST) Center has been conducting industrially relevant precompetitive research for over 30 years. As a collaborative effort between leading membrane industry sponsors and University researchers, the center focuses on developing innovative solutions to separations challenges. As an NSF and University Cooperative Research Center, MAST consists of four partner institutions: The University of Arkansas, The University of Colorado Boulder, The New Jersey Institute of Technology, and Pennsylvania State University. To learn more visit us at mastcenter.org.

A Millipore® mind
understands your world
before it designs for it.
Millipore.

NAMS 2022 – SPONSORS SILVER LEVEL

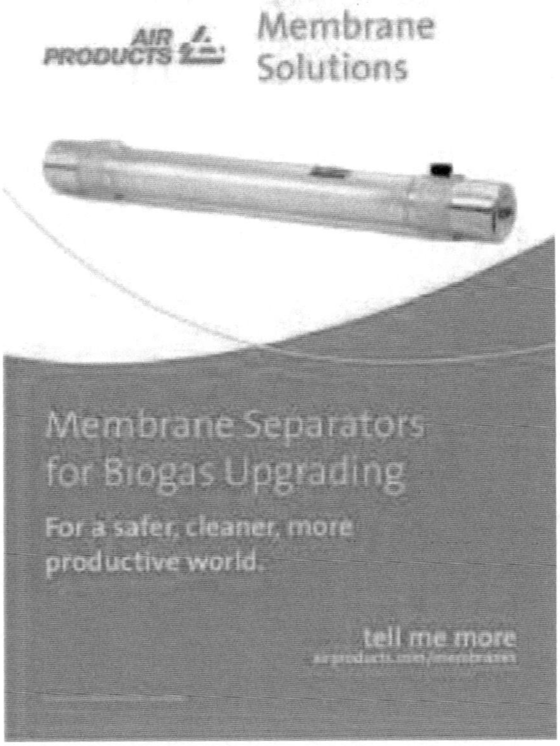

Chevron proudly supports NAMS

the human energy company

Honeywell
UOP

Membrane Solutions
for Industrial Separations

MTR
MEMBRANE
TECHNOLOGY & RESEARCH

ORAU

THE UNIVERSITY
OF ARIZONA

AMERIDIA
INNOVATIVE SOLUTIONS

EVONIK
Leading Beyond Chemistry

Gale & DB
Bhattacharyya

Georgia Tech ISSEC

The **Integrated Separations Science and Engineering Center at Georgia Tech** (ISSEC) is designed to foster collaborative & interdisciplinary solutions to separation challenges by promoting research partnerships not only among faculty from many units but also partnerships with industry. These collaborations results in new separation methods that are sustainable and frequently implemented by our industrial partners. Equally important, ISSEC helps support students and works with industry and government to learn the fundamentals of separations science and to apply them to real world applications.

NL Chemical Technology Inc.

www.nlchem.com

Expert

- Membrane for Water Treatment

Proprietary

- Technology (RO/NF) for Licensing

GORE

Together, improving life

Thanks to All Our Sponsors for Their Continued and Generous Support.

NAMS 2022 – EXHIBITORS

Anton Paar

NSF I/UCRC • Membrane Science, Engineering & Technology Center • MAST

Surface Measurement Systems
World Leader in Sorption Science

POROMETER

NAMS 2022 - SPONSORS

DIAMOND LEVEL

GOLD LEVEL

SILVER LEVEL

BRONZE LEVEL

NAMS

The North American Membrane Society

32nd Annual Meeting

May 13-17, 2023

THE UNIVERSITY OF ALABAMA®

Hosted by

Tuscaloosa, Alabama

Innovative sustainable membranes for water treatment and gas separation

NAMS 2023 will feature:
- workshops
- plenary speakers
- oral paper presentations
- poster session
- a banquet for conference participants and
- 5K run in Tuscaloosa!

Meeting Co-Chairs:

Milad Esfahani
mesfahani@eng.ua.edu

Steven Weinman
stweinman@eng.ua.edu

Jason Bara
jbara@eng.ua.edu

Follow us on Twitter: NAMS2023